PRENTICE-HALL FOUNDATIONS OF MODERN BIOLOGY SERIES

The Cell, *by Carl P. Swanson*

Cellular Physiology and Biochemistry, *by William D. McElroy*

Heredity, *by David Bonner*

Adaptation, *by Bruce Wallace and A.M. Srb*

Animal Growth and Development, *by Maurice Sussman*

Animal Physiology, *by Knut Schmidt-Nielsen*

Animal Diversity, *by Earl Hanson*

Animal Behavior, *by Vincent Dethier and Eliot Stellar*

The Life of the Green Plant, *by Arthur W. Galston*

The Plant Kingdom, *by Harold C. Bold*

Man In Nature, *by Marston Bates*

Cellular

Physiology

and Biochemistry

WILLIAM D. McELROY

The Johns Hopkins University

Prentice-Hall, Inc.

ENGLEWOOD CLIFFS, NEW JERSEY

1961

To Bill, Tom, Ann, Mary,
and Nell

Cellular Physiology and Biochemistry
William D. McElroy

Printed in the
United States of America
Library of Congress
Catalog Card Number 60–12211

PRENTICE-HALL FOUNDATIONS OF MODERN BIOLOGY SERIES
William D. McElroy and Carl P. Swanson, *Editors*

Design by Walter Behnke

Drawings by Felix Cooper

C

The science of biology today is *not* the same science of fifty, twenty-five, or even ten years ago. Today's accelerated pace of research, aided by new instruments, techniques, and points of view, imparts to biology a rapidly changing character as discoveries pile one on top of the other. All of us are aware, however, that each new and important discovery is not just a mere addition to our knowledge; it also throws our established beliefs into question, and forces us constantly to reappraise and often to reshape the foundations upon which biology rests. An adequate presentation of the dynamic state of modern biology is, therefore, a formidable task and a challenge worthy of our best teachers.

Foundations of Modern Biology Series

The authors of this series believe that a new approach to the organization of the subject matter of biology is urgently needed to meet this challenge, an approach that introduces the student to biology as a growing, active science, and that also *permits each teacher of biology to determine the level and the structure of his own course.* A single textbook cannot provide such flexibility, and it is the authors' strong conviction that these student needs and teacher prerogatives can best be met by a series of short, inexpensive, well-written, and well-illustrated books so planned as to encompass those areas of study central to an understanding of the content, state, and direction of modern biology. The FOUNDATIONS OF MODERN BIOLOGY SERIES represents the translation of these ideas into print, with each volume being complete in itself yet at the same time serving as an integral part of the series as a whole.

Contents

Introduction

To observe the dramatic properties of a single cell, we have only to peer through a small hand lens or a microscope. For centuries, biologists have marveled at these properties—at the shifting of the protoplasm inside the cell, at the changing size and shape of the cell, at the slow movements of cells away from or toward light. Below the reach of the microscope, however, lies a realm of smaller things, the world of the molecules and atoms that compose the ultimate structure of all matter. Biologists recognize the importance of the visible parts of the cell, but they are also aware that knowledge of the submicroscopic molecular pattern of the protoplasm is equally necessary to our comprehension of the structure and function of cells. In order to understand basic biological phenomena, therefore, the biologist today must be familiar with the latest developments in chemistry and physics.

The chief aims of biochemistry are to describe and analyze the chemical changes that occur in organisms. Investigations into the chemistry of living systems have shown that individual cells, whether of plants, animals, or microorganisms, are fundamentally similar in function despite vast differences in structure. Cellular physiologists are seeking to explain: the response of organisms and cells to their environment; the mechanism of cell growth, duplication, and reproduction; the ability of cells to take up nutrients from the environment; and the function and method of control of an organism's metabolic machine. With its discoveries so far, biochemistry, a science which is still in its infancy, has greatly illuminated the functional aspects of organisms as they affect the fields of agriculture, medicine, nutrition, and other older disciplines. The effectiveness of biochemical knowledge in solving practical problems depends on our mastery of the fundamental principles of the chemistry of life.

Biochemistry, therefore, involves more than an unraveling by organic chemists of the structures of the working parts of cells. Biochemists and cellular physiologists must make clear the relationship between structure and function in such a way that eventually we will be able to explain in chemical and physical terms many of the puzzling mysteries that still confront us: the nature of a cell's environment and its ability to resist and adapt to changes in that environment; the functional significance of the nucleus, of the mitochondria, with their array of enzymes, and of the microsomes and cell membranes that regulate the transport of materials into and out of the cell; the meaning of nutritional requirements and the importance of vitamins, inorganic ions, proteins, fats, and carbohydrates in cellular metabolism; and, finally, the ability of cells to utilize the stored or potential energy in chemicals and thus to enable organisms to move, speak, think, and reproduce.

These are some of the goals of biochemistry and cellular physiology. Although our knowledge has grown at an unprecedented rate during the past twenty years, much interesting and vital work remains to be done. In this brief book, we shall introduce you to the essential aspects of biochemistry and cellular physiology and discuss how they underlie the basic processes carried on by organisms.

1

The Chemistry of Cell Contents

All biological systems, whether of microorganism or of man, require certain chemical substances and certain chemical changes in order to stay alive, grow, and reproduce. The fundamental unit of these systems, the cell, must be able to take up nutrients from the surrounding fluid, and must contain the machinery for creating new parts of itself from the food material. Some of the nutrients taken up become part of the structure of the cell, while others are broken down to provide the energy needed to synthesize new molecules. In general, all these nutrients are called food, and include the carbohydrates, fats, proteins, minerals, vitamins, and water. Taken as a whole, these cellular processes are rather complicated, but if we examine the elaborate machine of the cell in the test tube, we can study the individual steps that lead to the synthesis of complex molecules.

During the past thirty years, biochemists have achieved remarkable success in isolating from cells or tissues specific parts of the metabolic machine. Once a particular chemical activity can be studied in the test tube, away from the complexities of the whole cell, rapid progress in the identification of the necessary factors involved usually results. In such studies, the biochemist draws on all the knowledge available from the fields of pure chemistry and physics; indeed, biochemistry can be described as the blending of chemistry, physics, mathematics, and related subjects in an effort to explain the physiological aspects

3

of cellular processes. As he has probed deeper into cellular function, the biochemist has discovered that certain groups of molecules are of unique importance to organisms. One of these groups consists of *enzymes,* which are the catalysts of chemical reactions in living cells. Enzymes are effective in very small amounts and are usually unchanged by the reaction they catalyze. The cell can duplicate itself in a few minutes or hours because its enzymes are capable of catalyzing the chemical changes associated with life processes. Since almost every individual chemical reaction in a cell is speeded up by a specific enzyme, we must know how these enzymes work. We shall defer a detailed discussion of enzymes until a later chapter. At this time we must investigate some of the chemistry of major cellular foods—proteins, carbohydrates, fats, vitamins, and other growth-promoting substances.

Proteins and Enzymes

Proteins are complex substances of high molecular weight which contain nitrogen in addition to carbon, hydrogen, and oxygen. The chemical and physical properties of enzymes clearly indicate that they are *proteins,* and when proteins are broken down in the presence of water (hydrolyzed) they yield a mixture of simple nitrogen-containing organic molecules called *amino acids.* The chemical composition of an amino acid is:

$$R-\underset{\underset{NH_2}{|}}{\overset{\overset{H}{|}}{C}}-C\overset{\displaystyle O}{\underset{\displaystyle OH}{}}$$

The $-NH_2$ group is the *amino group* while $-COOH$ is called the *acid* or *carboxyl group.* The R group can vary considerably; in fact, there are approximately 25 different R groups in nature and, therefore, 25 different amino acids. For instance, glutamic acid:

$$HOOC-CH_2CH_2\underset{\underset{NH_2}{|}}{C}-COOH$$

has an R group that is made up of two CH_2 groups and one carboxyl group. A few other amino acids are shown in Table 1. These are called *essential* amino acids because they cannot be manufactured by the body and therefore are indispensable in the diet of man. A person receiving a supply of nitrogen and carbon compounds as well as other nutrients can make the remaining *dispensable* amino acids.

Table 1

Essential Amino Acids

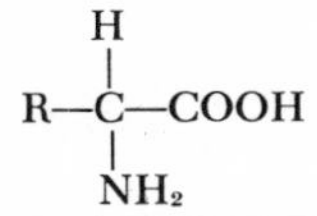

$$R-\underset{\underset{NH_2}{|}}{\overset{\overset{H}{|}}{C}}-COOH$$

Amino acid	R
Threonine	CH_3CHOH-
Phenylalanine	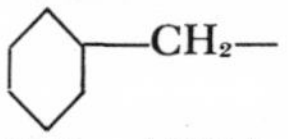 ⬡$-CH_2-$
Lysine	$NH_2-(CH_2)_3-CH_2-$
Tryptophan	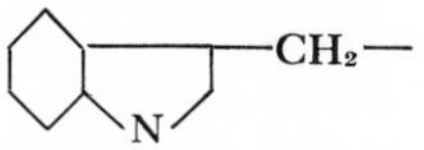 (indole)$-CH_2-$
Valine	$(CH_3)_2-CH-$
Methionine	$CH_3-S-CH_2CH_2-$
Leucine	$(CH_3)_2-CH-CH_2-$
Isolevcine	$CH_3-CH_2-CH(CH_3)-$

In the original protein molecule, the amino acids are joined together through bonds between the carboxyl group of one amino acid and the amino group (NH_2) of another by the removal of water (Fig. 1). This bond between two amino acids ($-CO-NH-$) is called the *peptide bond*. If only two amino acids are linked together, it is called a dipeptide, if three a tripeptide, etc., and if a large number are connected, it is a polypeptide. The long polypeptide chain of a protein can fold in a number of ways to make unusual shapes or configurations, and in many cases additional bonds (cross links) help stabilize the folded structure. As we shall see in Chapter 9, the folding of the long polypeptide chain is important in producing a stable and specific protein molecule, i.e., an enzyme.

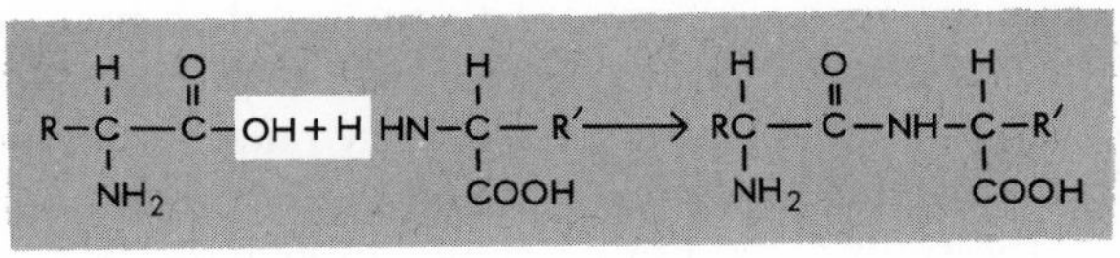

Fig. I. Synthesis of a peptide bond.

The sulfur-containing amino acid, cysteine, is a key member of many of these cross linkages. As shown in Fig. 2, when two sulfhydryl ($-SH$) groups are near one another, they can be linked together by oxidation, thus stiffening the folded polypeptide structure. The S—S link is called a *disulfide bond,* and it can be broken by substances which reduce it to the $-SH$ form. The S—S linkage is also important in joining

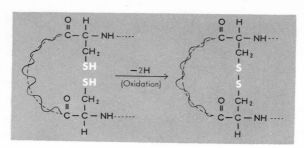

Fig. 2. The formation of a disulfide bond (cross link) in a polypeptide.

two or more polypeptides together. When we attempt to isolate and purify specific protein molecules from a cell, therefore, we should keep in mind that large aggregates or complexes of proteins can form by such cross linkages, in which case the apparent size of the protein molecules may be much larger than the basic unit we are interested in.

Enzymes consist of these large folded structures that apparently differ in number, kind, or sequence of amino acids present, or in the type and degree of folding. Why these unusually large molecules should act as catalysts remains an unsolved mystery of the cell, although we will present tentative explanations of their mechanism in Chapter 10.

In some instances, for an enzyme to function it must be associated with a small molecule called a *coenzyme*. In recent years, biochemists have made the significant discovery that parts of the coenzymes are often specific vitamins, such as riboflavin (B_2) and thiamin (B_1); we will discuss the function of coenzymes when we consider their need in metabolism. To function as catalysts, many enzymes also require specific metals such as iron, copper, or magnesium.

Because of the structure of proteins, and of the differences in coenzymes, the enzymes are usually highly specific; that is, a single enzyme will catalyze only one specific type of reaction, which means that the cell must produce a different enzyme for each chemical reaction. (There are some important exceptions to this statement, however.) It is sufficient for us at present to realize that each reaction will proceed in the cell at an appreciable rate only if the specific enzyme is present.

Naming enzymes is very easy once we know the type of reaction or the substance (substrate) acted on by the enzyme. Enzymes are denoted by the suffix, *-ase*. For example, an enzyme that catalyzes the breakdown of proteins is called a *proteinase*, one that catalyzes an oxidation is called an *oxidase*, and so on.

Carbohydrates

Sugars and starches are the principal sources of energy in the ordinary human diet, but they are not essential to the body or to its individual cells in any way, as far as we know. The cells could obtain their energy just as well from a mixture of protein or fat, or even from other types of carbon-containing molecules. Carbohydrates are the cheapest food commercially, and this is undoubtedly responsible for the large amounts contained in the average diet. Carbohydrates, although used primarily as an energy source, also supply carbon skeletons (carbon atoms linked together and associated with hydrogen and oxygen) that are necessary in the manufacture of the basic components of protoplasm. (For example, glucose, which contains six carbon atoms linked together, can be split to form two compounds containing three carbon atoms each. The three carbon units can then be used to make other essential compounds that contain only a three-carbon skeleton.)

The basic units in carbohydrates are composed of carbon, hydrogen, and oxygen. A few examples of carbohydrates, of varying chain lengths, are shown in Fig. 3. Note that in the specific examples (take glucose, for instance) the OH group can be written on the left or the right side. A three-dimensional model would show this much better, but the main point we wish to emphasize is that many sugars can contain the same number of carbon, hydrogen, and oxygen atoms and differ only in the position of the OH groups. These sugars of the same chemical composition but different chemical properties are called *isomers*. For example, besides glucose the sugars galactose, mannose, and fructose can be represented by the same formula, $C_6H_{12}O_6$.

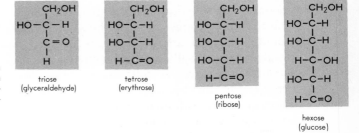

Fig. 3. Carbohydrates with different numbers of carbon units. Names of specific examples are given in parenthesis.

Since cells are able to discriminate between these isomers, because of the high degree of specificity of reaction shown by the enzyme proteins, the empirical formula of a compound does not give us information of any great value. In order to understand cellular metabolism, we must know the shape of the compounds, and thus we must consider molecular geometry when we discuss the reactions of all classes of compounds.

Sugars of the simple type can, like amino acids, be linked together to make longer units. Cane sugar (sucrose), for instance, is made up of a glucose linked to a fructose in a two-sugar chain that is called a *disaccharide*. Strings of three sugars, *trisaccharides*, are common, as are chains of many sugars, or *polysaccharides*. The polysaccharides such as starch and glycogen yield glucose when they are split with water (hydrolyzed).

Lipids

Proteins, carbohydrates, and lipids constitute the bulk of organic matter in cells. Lipids, or fats, are usually very insoluble in water, and as a consequence such organic solvents as ethanol or ether must be used to extract them from the cell. The lipids are a heterogeneous group of chemicals. The simplest ones contain only carbon, hydrogen, and oxygen, and by hydrolysis they yield glycerol and fatty acids. Most fatty acids are long straight chains in which the carbon atoms contain either the maximum number of hydrogen atoms (saturated) or a fewer number of hydrogens (unsaturated). The fatty acids may also have branched chains. A few examples are listed below; all contain the carboxyl (acid) group.

$CH_3CH_2CH_2COOH$—butyric acid—found in butter
$CH_3(CH_2)_6COOH$—octanoic acid—found in coconut oil
$CH_3CH=CH\ COOH$—crotonic acid—found in croton oil
$CH_3(CH_2-CH=CH(CH_2)_7COOH$—oleic acid—found in animal and
   plant fats

The fats are a concentrated food source, since they supply more than twice as many calories per gram as do carbohydrates and proteins. Cells are able to synthesize most, if not all, the fatty acids from the carbon skeleton of carbohydrates in the diet, and the few fatty acids that they cannot synthesize must be taken in with the diet, i.e., are essential. The amount of essential fatty acids required is small, and is provided by almost any diet. As we shall consider in greater detail later, the complex fats that are synthesized by cells are necessary not only because of their energy source, but because they form structural components of the cell. In particular, the cell membrane, as well as other submicroscopic particles in the cell that contain membranes, has a fat or lipid structure.

The fats (fatty oils such as olive oil and cod-liver oil) contain a mixture of what are called *triglycerides*, which are links between glycerol and fatty acids. As you will recall from chemistry, the reaction of an acid with an alcohol forms an *ester*. As shown in Fig. 4, the carboxyl group of the fatty acid reacts with the hydroxyl (OH) group of glycerol to form the triglyceride. The R^1, R^2, and R^3 compounds in Fig. 4 can be identical long-chain fatty acids (stearic acid, for example, would give the glyceride, tristearin) or they can be a mixture of fatty acids.

In contrast to the simple lipids are the waxes and related substances in which the glycerol is replaced by a longer-chain alcohol. Beeswax, for instance, is an ester of palmitic acid (a 16-carbon saturated fatty acid) and myricyl alcohol (a 30-carbon-chain saturated alcohol). In the simple

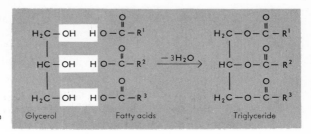

Fig. 4. The synthesis of a triglyceride.

lipid, one of the fatty acids may be replaced by compounds containing phosphorus and nitrogen to form the phosphatides, *lecithin* and *cephalin,* which frequently represent the major portion of cellular lipids. These compounds are soluble in both water and fats and therefore serve a vital role in the cell by binding water-soluble compounds (i.e., proteins) and lipid-soluble compounds together. Lecithin is a key structural material in the cell membrane, because it can maintain continuity between the aqueous and lipid phases of the inside and outside of the cell.

Minerals

A large number of inorganic mineral salts are essential for the growth of cells. The need for these minerals varies from cell to cell and particularly from plant to animal cells. They are often classified into two broad categories: *macronutrients,* which are required in large quantities, and *micronutrients,* which are required only in trace amounts. For both plants and animals, the major macronutrients are sodium, chlorine, potassium, calcium, phosphorus, and magnesium, and the principal micronutrients are iron, copper, manganese, and zinc. In addition to these, animals need cobalt, iodine, and possibly vanadium and selenium, and plants require boron, molybdenum, and vanadium, although vanadium is essential for only certain forms of plants. Most of the elements of the periodic table have been found in living cells, but this does not necessarily mean that all are essential to life. About 95 per cent of the ash of an organism is made up of potassium, phosphorus, calcium, magnesium, silicon, aluminum, sulfur, chlorine, and sodium, and the remaining 1 per cent or less is accounted for by the micronutrients and other elements. The micronutrients play a crucial catalytic role in enzyme systems. The other functions of minerals will be considered in Chapter 10 when we discuss the enzymes of cells.

Vitamins or Growth Factors

One of the most notable biochemical achievements during the past thirty years has been the discovery of vitamins and the analysis of their properties. Vitamins are relatively simple organic compounds, and although present in the body in very small amounts are absolutely essential to life. In this respect they are like the inorganic micronutrients discussed above. Vitamins are all different chemically, but none can be manufactured by the particular cells that require them. Since in an adult organism when an adequate amount of any vitamin is omitted from the diet a specific pathological condition or deficiency disease occurs, which is curable only by the administration of the specific vitamin, the vitamins are sometimes called *growth factors*. The other nutrients mentioned above are also essential for the growth and multiplication of the cell and could likewise be called growth factors, but most of them are required in larger quantities than are the vitamins. When vitamin-deficiency diseases were first studied, the chemical structure of these growth factors was not known, and they were therefore referred to as vitamins A, B, C, etc.

WATER-SOLUBLE B VITAMINS. When the B vitamins were first studied they were found to be soluble in water, in contrast to vitamin A, which is soluble in fat solvents. Other B vitamins include pyridoxine, nicotinic acid (niacin), pantothenic acid, biotin, folic acid, vitamin B_{12}, lipoic acid, and others whose status is not entirely clear. All water-soluble B vitamins are essential in cellular metabolism; at present we know of at least one enzyme-catalyzed reaction in which each of these vitamins functions.

FAT-SOLUBLE VITAMINS: A, D, E, AND K. Vitamin A exists only in animal products, but there is a yellowish substance in plants called carotene that can easily be changed into vitamin A by animal cells. Vitamin A is a fat-soluble substance (i.e., it dissolves more readily in fat solvents than in water) and has been shown to be essential for the growth of higher organisms and for the maintenance of normal nerve tissue as well as for the growth of bone and of tooth enamel. A deficiency of the vitamin produces night-blindness, the inability to see in a dim light.

Vitamin D is another fat-soluble vitamin, and since it is made in the body (from a plant substance called ergosterol that is present in the skin of animals which eat plants) under the influence of sunlight, it is sometimes called the sunshine vitamin. It is essential for the absorption of calcium from the intestinal tract. Vitamin E (alpha-tocopherol) appears to be necessary to prevent sterility in male animals; in cellular metabolism, it is involved to some extent in the process of oxidation of complex molecules such as carbohydrates and fats. Vitamin K also apparently plays a role in general electron transport or oxidative processes, and in adult mammals is essential for the normal coagulation of blood.

Gases

For the organisms that must live in air, oxygen itself can be considered as an essential nutrient. Carbon dioxide is usually classified as an end product of metabolism rather than as an essential nutrient, but in recent years studies have revealed that for bacteria carbon dioxide is an essential nutrient. Since carbon dioxide is normally produced in the metabolism of individual cells, we are not likely to observe a deficiency of this gas except under experimental conditions.

Biochemical investigations have discovered that many different types of molecules are necessary for the structure, maintenance, and function of the protoplasm of cells. Organisms vary considerably in their ability to manufacture these complex cellular components with their own metabolic machines. Some cells cannot make the required vitamins or amino acids from simple environmental materials and, consequently, these substances, as we have seen, become essential growth factors for such organisms. The study of growth requirements is called *nutrition* and encompasses all the methods by which food is utilized. Organisms are divided into two major categories, *autotrophic* and *heterotrophic organisms,* depending on their nutritional requirements. The nutrients needed by each type are a reflection of its internal metabolic machine.

Autotrophic Organisms

Autotrophs are organisms that are able to grow and multiply in a purely inorganic medium. From a nutritional viewpoint, they are the least exacting group of organisms, for they produce their own sugars, fats, amino acids, etc., from CO_2 and ammonia or nitrate. They obtain their energy in one of two ways and are subclassified on the basis of which type of energy they are dependent on.

CHEMOSYNTHETIC AUTOTROPHS. Organisms of this group synthesize all their protoplasmic constituents from CO_2, ammonia, or nitrate and obtain the energy for the synthesis by the oxidation of an inorganic substance. For example, two chemosynthetic autotrophs present in soil can oxidize ammonia and thus generate useful energy. The bacterium *Nitrosomonas* oxidizes ammonia to nitrite according to the following equation:

$$2NH_3 + 3O_2 \longrightarrow 2HNO_2 + 2H_2O + 79 \text{ calories}$$

Nitrobacter oxidizes nitrite to nitrate and thus receives the necessary energy for growth and multiplication. These two organisms are typical chemosynthetic autotrophs. No complex fats, carbohydrates, or nitrogen sources are required; only water, minerals, carbon dioxide, ammonia, and oxygen are essential for their functioning. Since the cells of these organisms contain all the complicated carbohydrates, fats, proteins, and vitamins usually found in cells, they represent magnificent synthetic factories for making all the complex components of normal protoplasm. Other chemosynthetic autotrophs (iron and sulfur oxidizers) have not been extensively analyzed and deserve further study by the cell physiologist and biochemist.

PHOTOSYNTHETIC AUTOTROPHS. These organisms produce the energy for their synthetic activities by converting light energy into chemical energy in the process of photosynthesis. They obtain their nitrogen from ammonia or nitrate and their carbon from CO_2. Organisms in this group are colored sulfur bacteria, blue-green, red, brown, and green algae, and complex green plants. The colors in the plants result from a mixture of pigments, including the crucial one, chlorophyll, which is capable of trapping light. Of all the biochemical processes in nature, *photosynthesis* is of paramount importance. By means of it, the green plants absorb the energy of sunlight and combine CO_2 with hydrogen (from H_2O, H_2S, or other sources) to build reserves of chemical energy (carbohydrates) that serve as the sole source of energy for all living things. The general equation for green-plant photosynthesis is:

$$CO_2 + 2H_2O \longrightarrow (CH_2O) + O_2 + H_2O$$

Water supplies the hydrogen for the reduction of CO_2 into the carbohydrate (CH_2O). Since the oxygen comes from the water, we show the interaction of two molecules of water on the left side of the equation even though a water molecule is regenerated in the process. In photosynthetic sulfur bacteria, H_2S is the hydrogen donor:

$$CO_2 + 2H_2S \longrightarrow (CH_2O) + 2S + H_2O$$

and in this case elementary sulfur is formed instead of oxygen.

Heterotrophic Organisms

The other major category of organisms is the heterotrophic group, composed of organisms that manufacture their energy mainly from organic sources, such as carbohydrates. Heterotrophic organisms, therefore, are related in their general metabolism to animals, while autotrophs are related primarily to plants. The heterotrophic organisms are divided into a number of subdivisions.

ORGANISMS THAT UTILIZE ATMOSPHERIC NITROGEN. A few species can trap atmospheric nitrogen and transform it, in a process called *nitrogen fixation,* into inorganic or organic nitrogenous compounds within the cells. Such organisms are invaluable in agriculture, since they form the natural fertilizers of the soil. A prominent member of this group is the bacterium *Azotobacter,* which is found in the soil. It can grow in the complete absence of other forms of nitrogen as long as it is provided with atmospheric nitrogen and a carbohydrate. Nitrogen fixation occurs in a large number of other organisms, but those of most interest are the root-nodule bacteria which grow on the small nodules of leguminous plants (clover, etc.). They can fix nitrogen only when they live in intimate (symbiotic) relationship with the plant. The leguminous plants are routinely used to enrich the soil with "fixed nitrogen."

ORGANISMS THAT DO NOT REQUIRE ORGANIC NITROGEN SOURCES. This group, one of the largest in the animal and plant kingdom, satisfies its nitrogen requirements by ammonia or nitrate; the carbon can come from a simple organic source, such as glucose or lactic acid, and the energy from the breakdown of any of a number of organic materials. The bacterium *Escherischia coli,* found normally in the intestine, is a typical example that can grow in a medium consisting of salts, including ammonia, and that draws on either glucose or lactic acid as a carbon and energy source.

ORGANISMS THAT REQUIRE CERTAIN AMINO ACIDS. The organisms belonging to the subgroups so far discussed synthesize all their amino acids from a source of inorganic nitrogen and a suitable source of carbon. Analysis of the proteins of these organisms indicates that all of the amino acids are present and that they must have been made, therefore, by the synthetic machinery of the cells. There are a number of organisms that require one or more specific amino acids for growth. In other words, they lack the machinery necessary for making these particular amino acids and are said to be *exacting toward* them.

ORGANISMS THAT ARE EXACTING TOWARD GROWTH FACTORS. Some species of organisms cannot synthesize certain vitamins or growth factors and must live, therefore, in an environment which can supply their nutrient. Certain microorganisms (bacteria and yeast) have been particularly useful in the study of such growth factors.

ORGANISMS THAT ARE EXACTING TOWARD BOTH AMINO ACIDS AND GROWTH FACTORS. Such organisms have ordinarily been nurtured for many generations in a medium in which all growth requirements are provided ready-made. If an organism's rate of growth is regulated by the rate of synthesis of some factor and that factor already exists in the environment, the organism will be able to grow more rapidly if it utilizes the pre-formed factor rather than wait for synthesis to occur. The synthetic process, therefore, may become altered or lost entirely. As we might expect, the richer the environment in growth factors, the poorer are the synthetic abilities of the organisms growing in it. Soil organisms have few or no complex growth factors supplied ready-made in their natural environment, but organisms that assume a parasitic existence in animal tissues are living in an environment rich in all those substances that go to make up protoplasm. In general, the more parasitic an organism is, the more exacting are its growth requirements, and some organisms are so dependent on several amino acids and growth factors that if any one of them is lacking in the medium, growth and reproduction are impossible. Thus parasitism, by increasing an organism's exacting growth requirements, forces it to become an even more eager parasite. Not all exacting organisms are parasites, however, or even pathogenic, since pathogenicity depends on factors other than parasitism. Organisms such as the lactobacilli from milk are highly exacting.

Man is a very complex, exacting heterotrophic organism. He requires an energy source, certain fatty acids, at least eight amino acids, several fat-soluble vitamins, and at least fifteen B vitamins. Recent studies of individual cells (liver, kidney, etc.) in tissue culture indicate that different organs have different nutritional requirements and that the over-all food requirements for man, therefore, cannot be taken as an index of the requirements of each cell.

We have not yet mentioned the most heterotrophic of organisms, the plant and animal viruses, and the bacterial viruses called bacteriophage, all of which are organisms that are incapable of growing and multiplying in the absence of other living cells. Most of them are composed of two chemicals, one a protein and the other a nucleic acid. They do not have a metabolic machine capable of reproducing themselves. The virus enters the host cell and carries along information in the nucleic acid that commands the machinery of the host cell to produce more of the virus. In some cases, the virus multiplies so vigorously that it destroys the host cell. These most exacting heterotrophs thus vitally influence the health of plants and animals and, in addition, offer scientists excellent material for studying the function of genetic material in an isolated system.

Although an organism may receive all the nutrients it needs, it will not grow and multiply unless other environmental conditions are favorable. Just as the nutritional requirements of organisms vary greatly, so do their reactions to the physical and chemical conditions in their environment. The most crucial factors in an environment are its oxygen concentration, acidity, alkalinity, and temperature, and we will now examine these influences to see how cells and organisms are affected by them. The response to a change in the environment often gives us some important information about the characteristics of the cellular machine.

Oxygen Requirements

Cells respond to oxygen concentration in one of three ways and are classified accordingly.

Strict aerobes. These organisms, which include the cells of all higher organisms except certain parasites, have an absolute requirement for oxygen. Certain cells of these higher forms can, however, go without oxygen for a short period. A muscle cell in our leg, for instance, can contract in the absence of oxygen although oxygen is needed to remove the metabolic wastes (lactic acid) produced under these conditions. Even though strict aerobes can function temporarily in the absence of oxygen, prolonged omission of oxygen eventually kills the cells or organisms.

Strict anaerobes. These organisms can grow and multiply only in the complete absence of oxygen, because oxygen, for reasons we will discuss later, is toxic to them. This group is composed primarily of bacteria.

Facultative anaerobes. These organisms live equally well with or without oxygen. For example, yeast cells under anaerobic conditions produce alcohol from glucose, and thus alcoholic fermentation (cellular activity in the absence of oxygen) is an expression of the cellular machine when it lacks oxygen. In the presence of oxygen, the yeast cell does not accumulate alcohol but metabolizes the carbohydrate completely to CO_2 and H_2O. Under both conditions the cells can grow and multiply.

During embryonic development, cells and tissues may actively shift from an anaerobic to an aerobic form of metabolism as they become more specialized. The mechanisms involved in this shift to aerobiosis are not understood, although a great deal of interesting experimental work is now being done on this problem.

Acid and Base

Some cells can withstand large variations in the acidity of their environment by preventing the changes in acidity from affecting the inside of the cell. An *acid* is a substance that forms hydrogen ions (protons) in solution. The concentration of the hydrogen ions determines the degree of acidity of the solution. A *base* is any substance that combines with hydrogen ions. Thus, in an aqueous solution, hydrochloric acid dissociates into hydrogen ions and chloride ions (a base):

$$HCl \rightleftharpoons H^+ + Cl^-$$

NaOH, on the other hand, dissociates into the metallic ion (Na^+) and a base (OH^-). Consequently, NaOH can neutralize the acid by forming water ($H^+ + OH^- \rightleftharpoons H_2O$) and the salt, NaCl.

As indicated above, water dissociates into equal numbers of H^+ and OH^- ions and is therefore neither an acid nor a base. At ordinary temperatures, the actual concentration of H^+ in pure water is very small, about 10^{-7} mols per liter. A molar solution contains one mole of solute per liter of solution. For example, if 58.5 grams of NaCl (the gram-formula weight of Na, 22.991, plus that of Cl, 35.457) are dissolved in water and made up to a volume of one liter, we have a 1-molar solution. If 5.85 grams are dissolved and diluted to a volume of 1 liter, we have a 0.1-molar solution, and so on.

The concentration of H^+ times the concentration of OH^- is a constant and equals 10^{-14} ($10^{-7} \times 10^{-7}$). If we know the concentration of one, then we can calculate the concentration of the other. Thus:

$$[H^+]\,[OH^-] = 10^{-14}$$

$$\therefore [H^+] = \frac{10^{-14}}{[OH^-]}$$

If a 0.01 normal HCl solution, for instance, completely dissociates into H^+ and Cl^- ions, the $[H^+]$ will be $0.01 = 1 \times 10^{-2}$ gm mols per liter, and the $[OH^-]$ will be 10^{-12} gm mols per liter. It is convenient to express $[H^+]$ and $[OH^-]$ in powers of 10 and to make use of logarithms; thus, $0.1N = 10^{-1}N$, $0.01 = 10^{-2}$, $0.001 = 10^{-3}$, etc. The exponents -1, -2, -3, etc., are the logarithms of the concentrations. Instead of negative numbers, however, we can make them positive by taking the negative logarithm. A hydrogen ion concentration of $0.001 = 10^{-3}$ gm mols per liter, therefore, has an acidity of 3, which is equal to $-(-3) = -\log[H^+]$. These numbers are called *pH values,* and they express the relative acidity of a solution; a pH value below 7 indicates an acid and one above 7 a base. Table 2 gives some examples of the relationship between $[H^+]$, $[OH^-]$, and pH.

Table 2

H^+ *Mols per liter*	OH^- *Mols per liter*	Log $[H^+]$	-log $[H^+]$ = pH
1×10^{0}	1×10^{-14}	0	0
1×10^{-1}	1×10^{-13}	-1	1
1×10^{-4}	1×10^{-10}	-4	4
1×10^{-7}	1×10^{-7}	-7	7
1×10^{-10}	1×10^{-4}	-10	10
1×10^{-13}	1×10^{-1}	-13	13
1×10^{-14}	1×10^{0}	-14	14

Most of the acidic and basic substances of biological interest do not dissociate completely. They are called "weak" acids or bases, and we may describe the behavior of a generalized weak acid (HA) in the following way: $HA \rightleftharpoons H^+ + A^-$.

Since the dissociation is not complete, the two kinds of ions and also the undissociated molecule will be in solution. The extent of the dissociation is expressed by the *dissociation constant* (K), which is defined as: $K = (H^+)(A^-)/(HA)$. We can rearrange this expression to read: $(H^+) = K \cdot [(HA)/(A^-)]$. If we take the reciprocal of this equation (that is, turn everything upside down), we get: $1/(H^+) = 1/K \cdot [(A^-)/(HA)]$. The logarithm of the above expression gives the following relationship: $\log[1/(H^+)] = \log(1/K) + \log[(A^-)/(HA)]$.

The $\log[1/(H^+)]$ has already been defined as pH. The $\log(1/K)$ is called the pK, and its significance may be appreciated if we consider what happens when this weak acid is titrated with some strong base such as NaOH. The reaction will form a salt (NaA), and the amount of this salt formed will be essentially equal to the amount of A^- that was present. We can substitute in the last expression and write: $pH = pK + \log[(salt)/(undissociated acid)]$.

To see what is happening in this titration of a weak acid, let us assume that the acid dissociates 1 per cent, which means that 99 per cent is still present as HA. If we add just enough NaOH to neutralize the 1 per cent H$^+$, thus removing the H$^+$, the remaining 99 per cent HA will dissociate 1 per cent again to form almost the same number of H$^+$, thus effectively keeping the hydrogen ion concentration constant. This is why these substances are called *buffers*.

From the final form of the above equation, we can see that when the second term on the right-hand side is equal to 0, the pH equals pK. This occurs when the acid has been neutralized 50 per cent. Then the numerator and denominator are equal and log 1.0 = 0.

As shown in Table 3, at 50 per cent neutralization large changes in acid alter the pH very little. Thus the greatest buffer action of a weak acid is observed at the pH that is equal to the pK of that acid. By selecting acids or bases with differing pK values (i.e., differing dissociation constants), therefore, we can find good buffers for different pH values. Phosphoric acid (H_3PO_4) is a good buffer for three different pH levels, because it has three different pK values (2.12, 7.20, and 12.66). Many compounds which occur in cells can act as buffers, because of their acidic or basic properties, and this fact is crucial since a change in pH will change the electric charge on large molecules such as enzymes.

Table 3

General Titration Data

Amount of base added (per cent)	Acid remaining (per cent)	$\dfrac{[salt]}{[acid]}$	$pH = pK + \log \dfrac{[salt]}{[acid]}$
0.1	99.9	$\dfrac{0.1}{99.9} = \sim 0.001$	$pH = pK - 3$
1.0	99	$\dfrac{1}{99} = \sim 0.01$	$pH = pK - 2$
10	90	$\dfrac{10}{90} = \sim 0.1$	$pH = pK - 1$
25	75	$\dfrac{25}{75} = 0.3$	$pH = pK - 4.7$
50	50	$\dfrac{50}{50} = 1$	$pH = pK - 0$
75	25	$\dfrac{75}{25} = 3$	$pH = pK + 0.47$
90	10	$\dfrac{90}{10} = \sim 10$	$pH = pK + 1$
99	1	$\dfrac{99}{1} = \sim 100$	$pH = pK + 2$
99.9	0.1	$\dfrac{99.9}{0.1} = \sim 1000$	$pH = pK + 3$

When we grow cells in culture, the pH should be maintained reasonably constant, and this is usually done through the addition of buffers to the growth medium. In complex organisms, the internal environment is kept constant by the operation of buffer systems that control the pH of the fluid around the cells. Thus in mammalian blood the pH is maintained near 7.35, and a number of inorganic as well as organic buffer systems help keep it there.

Not only the nutritional conditions, then, but the environment as well can alter metabolism. All enzymes are optimally active at a definite pH, and, consequently, as the environmental pH diverges from this optimum, the effectiveness of each enzyme unit decreases. The organism can sometimes compensate for this loss of efficiency per enzyme unit by producing more enzyme molecules. Those enzymes such as urease, catalase, and alcohol dehydrogenase, whose substrates are toxic to the organisms, generally respond in this manner. Remarkably, as far as we know there is no enzyme whose formation is not affected in some way or another during growth by the environmental pH. And some enzymes are largely formed only over a limited range of pH. As we shall see later, the ability of the cell to react to environmental changes is determined by the genetic constitution.

Temperature

Most animals and plants can exist only within a narrow range of temperatures—for the vast majority this range is between 10°C and 45°C. Some algae and bacteria, however, live in hot springs with temperatures as high as 90°C, and, at the other extreme, certain fish and invertebrates can carry on their life activities at temperatures of 0°C or even less. In general, the rates of chemical reactions in organisms increase with temperature up to a maximum and then decline as the temperature goes higher. Numerous studies have shown that this variation in the rate of chemical reaction with temperature is caused by changes in the enzymes. Enzymes have an optimum temperature at which they reach maximum catalytic activity. Above this temperature, the enzyme is inactivated, just as the protein of egg white is coagulated when the temperature is raised. At temperatures below the optimum, catalytic activity is decreased because of the slower kinetic activity of the molecules.

From this variation in enzymatic activity alone, we conclude that there must be certain optimum temperatures for growth and function. But since all enzymes do not have the same temperature optimum, no one temperature can be designated as the optimum one for growth and reproduction. If the first enzyme affected by a rising temperature is the one that regulates the synthesis of a cell nutrient, for instance, the supply of this nutrient ceases and growth stops at a relatively low temperature.

Permeability and the Osmotic Environment

When a cell is placed in a suitable nutrient medium, it increases in size and divides into two daughter cells that are apparently similar to the mother cell. Just because a nutrient is close at hand outside the organism, however, does not mean that the organism is capable of utilizing it, because the substance must first pass through the cell membrane into the machinery that synthesizes cellular constituents. All the cell's food and waste products must pass through this membrane, and to do this the penetrating substance must be soluble to a certain degree in the fluid around the cell or in the protoplasm itself. Since not all dissolved substances can penetrate the membrane with equal facility, the membrane is said to be selective, and this selectivity is vital in maintaining the life of cells. Although the cell membrane is the primary determinant of the cell's internal environment, the other intracellular units—the nucleus, mitochondria, microsomes, etc.—have selective membranes that control their own inner environment. A number of factors, including hormones, ionic environment, metabolic energy, pH, and temperature, affect the permeability of these membranes.

Whenever two different solutions are separated by a selectively permeable membrane, an *osmotic system* is established. Each cell, therefore, represents an osmotic unit, since the selective membrane always intervenes between the inner protoplasm and the external solution, and large quantities of water and lesser amounts of dissolved substances constantly pass into and out of the cell across the membrane. Before we consider the complex process of osmosis, let us examine the spontaneous migration of molecules within the limits of a single solution.

If any dissolved substance is concentrated more heavily in one part of a solution than in another, it will spread gradually until its molecules are evenly distributed throughout the whole solution. This process is called *diffusion,* and is caused by the random movements of all the molecules that make up the solution (Fig. 5). These movements are a manifestation of the kinetic energy of molecules, and the direction a chosen particle will take at any particular moment is entirely unpredictable, because its direction depends on its chance collision with other particles or with the wall of the container. Nevertheless, the *mass movement* of each kind of particle present in a solution can be predicted accurately on a statistical basis, for this movement is governed by the fundamental law of diffusion, which says that the particles of each different substance in a solution will diffuse from a region of great concentration to a region of less concentration. Diffusion will continue until every component reaches an equal concentration throughout the solution.

The relative concentration (the number of particles in a unit volume of solution) of each substance in solution is critical in determining the

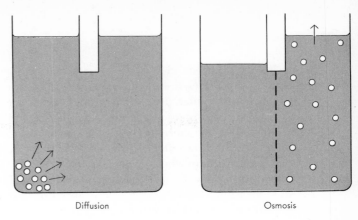

Fig. 5. Diffusion and osmosis.

Diffusion   Osmosis

direction of the diffusion of that substance. Since no two molecules can occupy the same space at the same time, an increase in the concentration of any one substance necessarily displaces an equivalent amount of all other components in the solution. Accordingly, whenever the total solute concentration is high, the solvent concentration must be low, and whenever the concentration of one solute is increased, the concentration of the other solute and/or the solvent must undergo a corresponding decrease.

The velocity of diffusion is determined by a number of factors. Since the whole process depends on the temperature, equilibrium is obtained more rapidly in warmer solutions, where the kinetic activity of the molecules is high. When the concentration difference is large, the particles have a greater tendency to escape from the concentrated region. Large particles diffuse more slowly than small ones, and the more viscous the medium the slower the diffusion. Equilibrium is reached very slowly when the distances involved are macroscopic, but within microscopic and ultra-microscopic limits, the concentration may become equalized almost instantaneously.

In biology, aqueous solutions predominate, and, consequently, osmosis may be defined as the exchange of water between the protoplasm and the solution that surrounds the cell. Take a simple osmotic system in which the solvent is water, which is separated by a selectively permeable membrane from a solution of sugar (see Fig. 5). In this case, both solute and solvent are under the same compulsion to diffuse, each away from the region of its own high concentration. In a perfect system, where the membrane prevents the sugar from moving, only the water is able to penetrate, and the two solutions can reach equilibrium only by the transfer of water from one solution to the other.

Since the water concentration outside the membrane is higher than that inside, water molecules will pass into the sugar solution and increase the volume of the water there until eventually the pressure of the column of water on the membrane compensates for the entrance of water. The pressure exerted by the solution is called the *osmotic pressure*. Earlier studies have indicated that a molar solution (one mole dissolved in a liter of solution) of sugar separated from water by such a membrane has an osmotic pressure of approximately 22.4 atmospheres at 0°C. This pressure can be related to the gas laws, since a gram molecular weight of a gas at atmospheric pressure occupies 22.4 liters, and if the gas is compressed to a volume of one liter it exerts a pressure of 22.4 atmospheres. We can thus calculate the osmotic pressure of a solution by using the following relationship: osmotic pressure $= CRT$, where C is the molar concentration, R is the gas constant (0.082), which we use to express the results in atmospheres, and T is the absolute temperature (273 $+$ C°). Thus the osmotic pressure for a 1-molar solution of sugar at 0°C is: O.P. $= 1 \times 0.082 \times 273 = 22.4$ atmospheres.

Since the gas laws were developed for a perfect gas, they do not hold exactly for solutions, but they are quite satisfactory for calculating an *isosmotic solution* for cells, i.e., a solution that has the same osmotic pressure as the cellular contents.

This relationship between concentration and osmotic pressure is valid only for non-electrolytes. For electrolytes (those compounds that can dissociate into two or more particles), the osmotic pressure is greater for a molar solution, since the pressure is determined by the number of particles. For example, if NaCl were 100 per cent dissociated in water, its osmotic pressure would be twice that of a molar solution of sucrose. In calculating the osmotic pressure of an electrolyte solution, therefore, we must multiply by the degree of dissociation.

The cell quite obviously is not a perfect osmotic system. Not only water but many dissolved substances commonly present in and around the protoplasm are able to penetrate the cell membrane in significant amounts, so that the exchange of water between the cell and its surroundings is accompanied by the exchange of other substances. Oxygen and carbon dioxide readily pass into and out of the cell. The permeability of the complex cell membrane depends not only on the nature of the surrounding particles, but also on the changing conditions inside and outside the cell.

Although permeability varies in different cells and sometimes on different sides of the same cell membrane, certain generalizations can be set forth. For example, we know that water penetrates most cells rapidly. Gases such as carbon dioxide, oxygen, and nitrogen and fat-solvent compounds such as alcohol, ether, and chloroform easily penetrate all cell membranes. Somewhat slow to penetrate are such organic substances as glucose, amino acids, glycerol, fatty acids, etc., and slower still are the strong electrolytes—the inorganic salts, acids, and bases, and the large molecules of the disaccharides: sucrose, maltose, and lactose. Some cells can take up very large and complex compounds but almost none can absorb proteins, polysaccharides, or phospholipids.

There are many exceptions to these general statements. The bulk of all solutes present in protoplasm, including proteins and most sugars and inorganic salts, penetrate the cell membrane very slowly if at all. The solvent water, on the other hand, enters and leaves the cell very quickly and this, coupled with the fact that water is more abundant than all other components combined, means that the water must bear the main burden of establishing an osmotic equilibrium between the cell and surrounding solutions. If the cell is placed in a solution with a water concentration drastically different from that in the protoplasm, so much water will enter or leave the cell that it may be destroyed. The cell wall in plants and in microorganisms is usually rigid enough to prevent the swelling of the cell when water rushes in.

In an isosmotic solution, the concentration of water is the same as that in the protoplasm, and this condition occurs only when the total concentration of solute particles in the solution equals that in the protoplasm. This water balance is achieved because the water molecules that are continuously escaping from the cell are matched by an equal number entering the cell. A true isosmotic solution, therefore, contains a concentration of non-penetrating or very slowly penetrating solute molecules that approximates the total concentration of non-penetrating solutes in the protoplasm. An equal water concentration inside and outside the cell could not otherwise exist.

The best results in preparing an isosmotic solution are obtained by including salt mixtures in which sodium, potassium, calcium, and magnesium ions are represented in the proper proportions, since these ions are often essential to maintain the normal permeability of the cell membrane. A *hypotonic solution* contains a relatively low concentration of non-penetrating solutes, compared to the protoplasm of the cell it surrounds, and thus the water concentration in it is relatively high. Cells placed in hypotonic solutions, then, tend to take in water and swell, in accordance with the fundamental laws of osmosis and diffusion. If the solute on the outside is very low, the swelling will continue until the cell membrane ruptures. When human red cells are placed in a solution containing only 0.2 per cent sodium chloride instead of the usual 0.9 per cent present in an isosmotic solution, the corpuscles swell and burst even before they can be observed under a microscope.

Cells placed in a *hypertonic solution,* on the other hand, tend to shrink, because the solution contains a higher concentration of solute molecules than does the protoplasm. In plant cells, the rigid wall maintains the original form of the cell, and the protoplasm draws away from the wall. Animal cells, however, when placed in a hypertonic solution shrivel up so much we cannot detect the original shape of the cell. When a red blood cell, for example, is placed in a concentrated sucrose solution, it shrinks from loss of water. If this process is not carried too far, it is reversible. Thus if we place a plant cell in a hypertonic solution whose solute particles will slowly penetrate, the plant membrane will contract away from the cell wall (*plasmolysis*), but then as the solute particles slowly enter, water will return and the cell will resume its original size (*deplasmolysis*) (Fig. 6). Observing deplasmolysis, we can determine the rate of penetration of solute particles into plant cells, for the rate of deplasmolysis obviously parallels the rate of penetration by the solute particles.

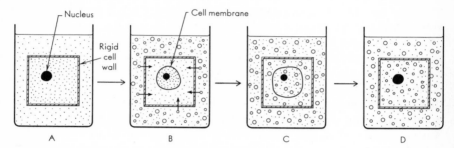

Fig. 6. Plasmolysis. When a plant cell (A) is placed in a hypertonic solution of a slowly penetrating solute (B), water is rapidly lost from the cell, and the cell membrane shrinks away from the cell wall. As the solute slowly penetrates (C and D), water also re-enters, and the cell swells to resume its original size (D).

If we remove the rigid cell wall from a bacterium or plant cell, we find that, like an animal cell, it is no longer able to maintain its shape. In a hypotonic solution, for example, the cell will burst. Normally, however, water enters the plant cells from a hypotonic solution until the proto-plasm of the plant cell is forced outward against the unyielding cell wall. When a sufficiently high internal pressure (*turgor pressure*) is generated, often amounting to several atmospheres, no more water can enter.

By placing the cells in varying concentrations of a sugar solution, we can estimate the number of osmotically active particles inside the cell. If we immerse a plant cell in a 0.5-molar solution of sugar and the cell neither shrinks nor swells, we know the cell contains the equivalent of 0.5 molar osmotically active particles (both electrolytes and non-electrolytes), which would, in effect, exert a pressure of 12 atmospheres. Many plant cells, including bacteria, have this concentration of osmotically active particles inside. Cells such as bacteria that have their cell walls removed will thus swell and burst when placed in water, because they are unable to withstand the 12 atmospheres of pressure. Penicillin kills growing bacteria by preventing the synthesis of a new cell wall, thereby exposing them to the hypotonic solution that literally blows them up. To maintain these bacteria (called *protoplasts*) that have no cell wall, we increase the concentration of the solute—usually 0.4 to 0.5 molar sucrose.

We do not want to leave the impression that the selective membrane is passive and that simple diffusion and osmosis account for all exchanges between cells and the surrounding fluids, for many cells can accumulate certain substances and exclude others against the natural tides of diffusion. Many marine algae, for instance, accumulate iodine to a concentration which is more than a million times greater than that of the sea. The cytoplasm of cells generally is exceedingly rich in potassium and poor in sodium compared to the surrounding sap or lymph, and this situation cannot be accounted for by the simple laws of diffusion or osmosis. The cell, therefore, often spends energy to force the molecules of a particular substance to move against a concentration gradient (active transport). Since the energy to generate these forces is derived from the metabolic process, if metabolism is temporarily suspended, the cell loses its capacity to work against the tide, and the laws of simple diffusion and osmosis will control all exchanges. For identification, we customarily designate the moving machinery as "pumps," and thus speak of the sodium and the potassium pump.

We now know specific structures in the cell membrane regulate the pumping of substances into the cell. The units on the cell membrane are under genetic control and can be eliminated by a change in a specific gene. As we shall see later, a cell may contain an enzyme that catalyzes the metabolism of a particular substance but is unable to utilize the external supply of this substance because it cannot transport the substance through the cell membrane. The enzyme, therefore, can attack only those molecules that are produced internally. By gene mutation and selection, it is often possible to reverse this process and obtain a cell type which has a new specific pump. It is evident that the cell membrane is not a passive selective permeable barrier, but rather an active partner of the metabolic machine, with many of the properties we normally ascribe to enzymes.

When nutrients pass through the cell membrane, they enter a new environment: the metabolic machine of the cell. This machine's primary purpose is to convert these nutrients into useful energy and at the same time to create new carbon skeletons that are needed to synthesize other vital compounds. For the past seventy-five years, research into the cell's metabolism has been a primary activity in biochemistry, and we have now gained a reasonably good understanding of the various reactions in metabolism. Before considering these reactions in detail, however, we must see how energy is liberated in chemical reactions.

For atoms to be joined by specific bonds into molecules, work must be done, i.e., an external source of energy is needed to bring the two atoms together to form a stable molecule. In nature a prime example of energy conversion is sunlight's ability to "put" CO_2 and hydrogen together into high-energy carbohydrates. The energy in the light rays, in effect, is used to make specific chemical bonds. When these bonds are broken, the energy that holds the atoms together is liberated and may be employed to do work. Thus, when an organism, during metabolism, breaks or splits certain atomic links, we want to know whether energy is liberated and whether this energy is released in a form capable of doing useful biological work.

When substance A is converted to substance B, therefore, we must be able to determine whether energy is absorbed

4

Metabolic
Energy

or liberated. Suppose we set the energy level of *A* at some arbitrary value. If energy is given off when *A* is converted to *B*, the energy level of *B* is lower than that of *A*, and we say it has a minus value. But if energy is absorbed when *A* is converted to *B*, then *B* is said to be energetically more positive than *A*. If energy is given off, we want to determine whether all or some of it is available to do work on another system. We call this "employable" energy *free energy*. In biochemical systems this free energy is not literally liberated into the environment, but is conserved as special *bond energy* that is passed on to other molecules and used there to form new bonds.

Free energy has two essential characteristics, which can be easily understood if we consider the energy that can be obtained from water pouring over a dam. The first characteristic is the *distance* the water falls, the second the *quantity* of water that spills over the dam. The product of the two factors equals the total amount of work that can be performed. In his classical studies, Joule let water fall on a paddle from a definite height and measured the temperature change of the water. From this experiment he was able to define the relationship between calories and the mechanical equivalent of heat. A *calorie* is the quantity of heat required to raise the temperature of one gram of water one degree centigrade, starting at 14.5°C (15° calorie) and is equal to 4.185 joules.

Returning to our dam, we see immediately that just a bucket of water will do very little work, even if the distance the water drops is considerable. It takes a sustained flow, and thus a large reservoir above the dam is needed to perform an appreciable amount of work. In other words, a system low in water reserves (or energy potential) has very little capacity to do work.

Let us examine a simple water reservoir as an example (Fig. 7). If we open the gate, the water flow from *A* to *B* will accomplish work until it is equalized by the flow from *B* to *A*. When this reversible process has reached equilibrium, no more work can be gained and the net free-energy change is zero. In a chemical reaction, say when *A* goes to *B*, if we know the amount of *A* that is converted into *B* up to the time of equilibrium, we can calculate the free energy that has been liberated.

Fig. 7. When water flows from one reservoir to another (*A*→*B*), work can be performed. The amount of energy available depends on the amount of water and how far it falls. At equilibrium the flow from *A* to *B* equals the flow from *B* to *A*, and no net work is possible.

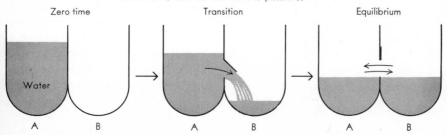

Thus a relationship exists between the free energy liberated in a reaction and the extent of the reaction, and we calculate it with the aid of what we call the *equilibrium constant,* which is the ratio of the product, B, and the reactant, A, after the reaction has reached equilibrium. We have then:

$$\Delta F = -RT \ln K$$

where K is the equilibrium constant (B/A), R is a constant (approximately 2) that is necessary to convert the units into calories, and T is the absolute temperature. If logarithm to the base 10 is used, we multiply the quantities on the right by 2.3 and get:

$$\Delta F = -4.6\ T \log K$$

If at equilibrium most of A was converted to B, K would be large and ΔF would be large and negative; the negative sign indicates that energy was liberated. If very little A is converted to B, the equilibrium constant may be less than 1, and ΔF would be positive, indicating that the reaction favors the formation of A. When energy is liberated in a chemical reaction ($-\Delta F$), we call it an *exergonic* process. If energy must be supplied to promote the reaction ($+\Delta F$), we call it an *endergonic reaction.*

The equilibrium constant is an important quantity, for it tells us something about the energy levels of the reactants and products. We may not be able to say precisely how the energy is liberated and possibly transferred to other systems, but we do know from the equilibrium constant what to expect energetically in the bond-breaking process.

In some cases we may not be able to measure the equilibrium constant, and we then have to depend on other methods for determining the energy that is released in a chemical reaction. This is often true for oxidation and reduction reactions. Biologically significant free-energy changes occur in oxidation and reduction reactions, and in these the changes are related to the electrical equivalent of work. In oxidation, electrons are removed; reduction is the reverse process. Since the movement of electrons from one energy level to another (or from one compound to another) is similar to the electrical current in a wire, work can be done by biological oxidations. All we need to know to calculate the free-energy change involved in electron movement is the change in the potential-energy level of the electron.

By studying the capacity of compounds to take up electrons (their oxidizing capacity), we can gain some notion about how they compare as oxidizing or reducing agents. We could, for example, by placing a metal electrode in a solution of a substance, determine whether the substance in question passed electrons to the electrode. If this were the case, we would say that the substance had a higher *electron pressure* (greater reducing power) than the electrode. By taking an electrode and arbitrarily assigning it an electron potential of zero, we can calculate in arbitrary units the oxidizing and reducing strength of a number of compounds of biological interest. From such measurements we can tell how compounds will act relative to one another as oxidants or reductants. When two such systems are coupled together, we can determine the amount of energy liberated or absorbed in the reaction if we know the difference (ΔE) in the potential between the two. This situation is analogous to the water-reservoir system, except that now we use electrons instead of water. The free-energy change produced by the electron flow is calculated from this relationship:

$$\Delta F = n\ 23{,}000\ \Delta E$$

where n is the number of electrons that move and ΔE is the change in the electrical potential. The constant, 23,000, is necessary to convert the units into calories.

There is one other important way to calculate the free-energy change in a chemical reaction, and it leads us into a discussion of the true meaning of free energy. From our discussion thus far, it is obvious that free energy is only that energy available to do useful work and is different from the total energy change of a reaction. Thus:

total energy change = utilizable energy change + non-utilizable energy change

This non-utilizable energy change may be small or large, depending on the system, and is a reflection of the change in the ordered state of the system called the entropy (S) change. The more ordered or improbable the state is, the less entropy there is, and the more random or probable the state is, the more entropy. In other words, the entropy of a system tells us how much the system is run down. The non-utilizable energy change is then expressed as the absolute temperature (T) times the change in entropy (ΔS) in going from one state to another. The above expression can be rewritten as follows:

$$\Delta H = \Delta F + T\Delta S$$

Usually this is turned around to show the relationship of free energy to the other quantities:

$$\Delta F = \Delta H - T\Delta S$$

We designate ΔH as the change in heat that results in going from one state to another. If heat is given off, ΔH is negative; if heat is absorbed, ΔH is positive. If we study the reaction, $A \rightleftarrows B$ at 27°C (300°

absolute), we can determine ΔH by placing A and B in a bomb calorimeter and determine the amount of heat given off by each upon combustion to CO_2 and water. The difference in the heat liberated is ΔH. If we assume 5000 calories were liberated from the combustion of A and only 2000 from B, then $\Delta H = -3000$ calories. If the entropy change in going from A to B was 10, we can calculate ΔF as follows:

$$\Delta F = -3000 - 300 \times 10 = -6000$$

Note that ΔF is negative, i.e., energy is given off, and that the entropy change makes a significant contribution to the ΔF value.

Oxidation and Energy

The oxidation of carbohydrates or related compounds is the main source of energy for many organisms. Since organisms are not heat engines, they must liberate the energy in a utilizable form; specifically, they must trap this energy from exergonic processes. Unfortunately, free energy is often considered to be a tangible object, a package of energy that can move from one place to another. Actually, in chemical reactions the energy is first redistributed in the molecule in such a way that when additional molecular bonds are broken, energy is liberated to the environment as heat. Thus organisms must trap the energy-rich intermediate and use it before the energy is lost as heat. An "energy-liberating reaction," therefore, generates an "energy-rich group," which is very reactive and is capable of initiating reactions requiring energy. Energy-yielding reactions must be coupled with energy-consuming reactions in an intimate way if energy is not to be lost as heat. Under these circumstances, the energy is not "liberated" as such but is redistributed in the reacting molecules. In many of these coupled reactions, the over-all process proceeds spontaneously.

The oxidative reactions of the cells are the ones that most actively redistribute or "liberate" energy. Although workers have applied the term oxidation only to reactions in which oxygen combines with another substance, to be perfectly accurate oxidation is any process in which hydrogens or electrons are removed, for oxygen is just one of many biological hydrogen-acceptors.

The simplest type of oxidation is called dehydrogenation. Hydrogen is removed, as occurs in the oxidation of alcohol, $RCH_2OH \rightarrow RCHO + 2H$, or electrons are lost, as is the case in the oxidation of iron, $Fe^{++} \rightarrow Fe^{+++} + $ electron. Oxidations and reductions cannot proceed, of course, by these half reactions; every substance undergoing oxidation must be accompanied by a substance undergoing reduction, and vice versa. Oxygen itself can be the oxidant and, when reduced, forms either water or hydrogen peroxide (H_2O_2). The substances that take part in these reactions have a tendency either to eject or attract electrons, and this tendency is directly proportional to their strength as reducing or oxidizing agents. This is another way of stating what we have already said about the free energy of a reaction: When an electron flows from a reducing system to an oxidizing system, energy is liberated. But exactly how is energy liberated and trapped in an oxidation-reduction reaction?

First let us consider a reaction, the oxidation of an aldehyde. We will assume for the moment that we understand how enzymes act, since enzymes are necessary in trapping the energy in the catalytic oxidation of an aldehyde. Enzymes often contain an SH group (sulfhydryl), as is indicated in Fig. 8. The aldehyde first combines with the enzyme by splitting the double-bond oxygen in the aldehyde part of the molecule, thus producing a compound that contains an OH group and a single hydrogen on the terminal carbon. One of the double bonds to the oxygen connects with the sulfur in the enzyme molecule, and the hydrogen normally associated with the sulfur moves to the oxygen to satisfy its valence.

In the next step the oxidation process occurs, and energy in the molecule is redistributed (it does not "liberate" energy as is sometimes stated). The departing hydrogens or electrons are accepted by another compound, which we will merely call X, to form XH_2. Since this X, or oxidizing agent, is at a lower energy level than the initial aldehyde compound, when electrons from the aldehyde flow to it they deplete the number of electrons in the outer shell of the aldehyde atoms, enabling the atoms to occupy a different energy level. This nuclear rearrangement, which we usually call energy liberation, results because the nuclei seek

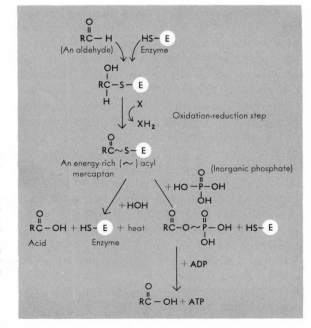

Fig. 8. The oxidation of an aldehyde to an acid and the generation of an energy-rich group. Note that in the oxidation of the aldehyde, an energy-rich ($\sim$) intermediate is formed on the enzyme. If water splits this intermediate, the energy is lost as heat, whereas if phosphate splits the intermediate, the energy is conserved in the phosphate bond and can be stored in the form of ATP.

lower positions of potential energy than they occupied when they held the two electrons around them; in going to a lower energy level in oxidation, they redistribute energy in the molecule. Note that the product of the oxidation contains a double-bond oxygen and is linked to the sulfur atom in an arrangement called "an energy-rich bond."

Energy-rich Bonds

If the link between the carbon and sulfur is broken by water (hydrolysis), a large amount of energy in the form of heat will be liberated. When the molecule is hydrolyzed, it forms an acid (carboxyl group) and the SH enzyme once again. In short, the aldehyde is oxidized in the presence of water to form an acid and a reduced substance. On the other hand, if the organism is to utilize this energy in the bond it must transfer the energy-rich group to some other molecule, thus synthesizing a new compound by drawing on the energy of the carbon-sulfur link. If the organism does not immediately need the reactive group for synthesis, it can store the energy in some other chemical form, thus freeing the enzyme and conserving the bond energy in a reservoir. *The coupling of this bond energy to other systems is the key reaction in biosynthetic processes.* Instead of hydrolyzing the oxidation product, which is called an acyl mercaptan, we can split it with phosphorus acid or phosphate (H_3PO_4). The phosphate group prevents the loss of energy of the link, resulting in the formation of an "energy-rich" link between the carbon and phosphorus.

Another important phosphorylated compound, adenosine diphosphate (ADP), whose structure is shown in Fig. 9, will react with this energy-rich phosphate group. The product of the reaction is adenosine triphosphate (ATP). ATP has proved to be universally distributed in plants, microorganisms, and animals, and it serves as the initial storehouse for the energy generated in oxidative reactions. This bond energy of the phosphate groups directly or indirectly drives all the energy-requiring processes of life.

Thus the energy of an oxidation reaction creates an energy-rich group, and, as we shall see later in our discussion of metabolic processes, this bond energy must couple many reactions that require energy before they can proceed to any significant degree. It turns out that even the second phosphate in ATP is energy-rich, but the third one, which is on the number 5 carbon of the pentose, is not energy-rich. ATP, therefore, is capable of supplying two energy-rich units. The link between two energy-rich phosphates, as in ATP, is called a *pyrophosphate bond,* and has proven to be the primary grouping in the conservation of oxidation-reduction energy. We shall discuss numerous examples later when we take up the way the pyrophosphate bond energy of ATP is used in biological reactions.

When ATP is hydrolyzed to liberate the two terminal phosphates,

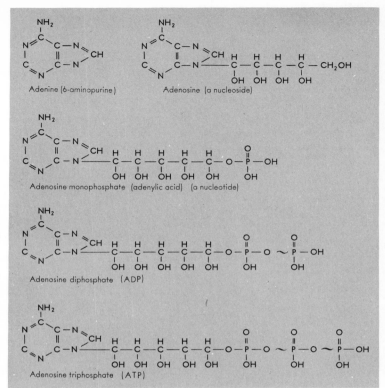

a large amount of heat is liberated—at pH 7.0 the free energy of hydrolysis of the terminal pyrophosphate bond in ATP is approximately 8000 calories, in contrast to the free energy of hydrolysis of an ordinary ester, which may be no more than 1000 or 2000 calories.

Instead of talking about the free energy of a biochemical reaction being *liberated,* therefore, we speak in terms of the creation of bond energy that can be used to drive other reactions. We recognize an energy-rich bond by its ability to couple with other energy-requiring processes. When an aldehyde is oxidized to an acid, the energy does not come off as little packages but is conserved in the form of bond energy. The other way that energy can be redistributed in a molecule so that it can generate an energy-rich phosphate group is by the removal of water; we will study this method when we consider carbohydrate metabolism.

In this discussion of energy metabolism, we do not want to leave the impression that energy is the only significant factor in a given reaction. In the metabolism of a cell, certain key reactions lead to the formation of molecules that are necessary for physiological functions. Although some of these processes may be part of an energy-yielding reaction sequence, the structure of the product may be of considerable importance in initiating other synthetic reactions.

5

Alcoholic Fermentation

Although man has been familiar with alcoholic fermentation since prehistoric times, what causes it was not discovered until about 1860. A short while before this, however, the French chemist, Gay-Lussac, had succeeded in describing the production of alcohol from sugar by this equation:

$$C_6H_{12}O_6 \longrightarrow 2CO_2 + 2CH_3CH_2OH$$

In the middle of the nineteenth century, two schools of controversy developed concerning the mechanism of alcoholic fermentation. One group considered the process to be strictly a chemical one, while the other claimed that it was a process intimately associated with the activities of living organisms. During this controversy, Pasteur was studying how lactic acid is produced in milk and how acetic acid is formed in certain wines. From his experiments he concluded that fermentation takes place anaerobically only in the presence of certain microorganisms; when these organisms were excluded no fermentation occurred, thus indicating that fermentation is a physiological process that is closely bound up with the life of cells—an idea, of course, that contradicted Lavoisier's earlier conclusion that all living things need oxygen for respiration. In challenging Lavoisier's views, Pasteur suggested that there were other substitutes for molecular oxygen and that fermentation, indeed, is the result of life activities being carried on in the absence of air. The experiments and reports of Pasteur did much to establish the belief that fermentation is directly associated with the growth and metabolism of living matter.

More than twenty years elapsed before the next major breakthrough in the investigation of fermentation. In 1897 Buchner accidentally stumbled on a discovery that not only opened the door to the secrets of fermentation, but to those of the whole of modern enzyme chemistry as well. Primarily interested in making what are called protoplasm extracts from yeast to be injected into animals, he ground yeast with sand, mixed it with certain other compounds, and finally squeezed out the juice with a hydraulic press. Since it was difficult to prepare this material each day, he made various attempts to preserve the cell-free extract. Because it was to be injected into animals, ordinary antiseptics could not be used, so he tried the usual kitchen chemistry of adding large amounts of sugar.

When the sucrose was rapidly fermented by the yeast juice, he observed fermentation in the complete absence of living cells for the first time. At last it was possible to study the processes of alcoholic fermentation independently of all other processes such as growth, multiplication, excretion, and the additional complications that accompany fermentation in the living yeast cell. Buchner's work was soon followed by intensive studies of the properties of yeast juice, which was found to be capable of fermenting a large number of sugars such as glucose, fructose, mannose, sucrose, and maltose. Glucose was converted by the juice into ethyl alcohol and carbon dioxide, according to the Gay-Lussac equation.

The first important analysis of the activity of yeast juice was made by Harden and Young in 1905. When they added fresh yeast juice to a solution of glucose at pH 5, fermentation began almost at once. Although the rate of carbon dioxide production soon fell off, they restored it by introducing inorganic phosphate. The recovery was only temporary, however, because the phosphate was soon used up, and the rate of fermentation dropped as the phosphate concentration declined. By adding more phosphate, they sparked another burst of fermentation. Since the inorganic phosphate they added to fermentation mixtures always disappeared, the two scientists suspected that an organic phosphate ester was being formed, and they soon confirmed their suspicions by isolating such an ester in the form of a hexose diphosphate (fructose diphosphate). They observed that this substance was fermented very actively by fresh yeast juice (it was like glucose in this respect) and concluded that it was probably an intermediate in alcoholic fermentation.

Later Robison isolated another sugar phosphate. When examined in detail, the sugar phosphate proved to consist of an equilibrium mixture between glucose-6-phosphate and fructose-6-phosphate. Since both of these sugar phosphates could be fermented, it seemed clear that these substances must arise from a coupling of inorganic phosphate with glucose on carbon number 6. How these esters are formed and in what way the fructose diphosphate is eventually converted into alcohol and carbon dioxide were questions that stumped a number of brilliant biochemists for decades. As we shall see later, studies of muscle extract revealed some of the chemical steps in fermentation, and it gradually became apparent that the fermentation of glucose by yeast juice is very similar to the anaerobic breakdown of glycogen by muscle extracts.

We return to Harden and Young for the next stride forward in our understanding of fermentation. They discovered that yeast juice loses its activity when it is dialyzed, i.e., when the juice is placed in a selective permeable bag surrounded with water. The bag was much like a cellophane membrane, and the tiny holes in it permitted small molecules to move out into the water while the larger molecules stayed inside. Testing the contents of the bag, they found that the dialyzed juice would not ferment but that the activity could be restored by adding the material which had diffused into the water. Additional experiments demonstrated that the contents of the bag were easily destroyed by heating but that the material outside the bag was relatively stable.

From these important experiments, the co-discoverers found that in addition to the large molecules inside the bag, which we now know are enzymes, the yeast juice also contained dialyzable, thermostable substances which are necessary for the function of the enzymes. These dialyzable factors that move away from the catalysts inside the bag are the coenzymes. Thus yeast juice was thought to contain a zymase, which is composed of a non-dialyzable thermolabile enzyme, and a cozymase (cofactor), which is dialyzable and thermostable. We now realize, of course, that the zymase is a complex mixture of many enzymes and that several coenzymes are necessary for their function.

At this point we are going to make a big jump in the history of the development of carbohydrate metabolism and consider our present knowledge of the initial steps in glucose metabolism. It is neither necessary nor desirable to give all the details here about the intermediates in the conversion of glucose to alcohol. They are contained in advanced texts and monographs, and some of the prominent ones are listed at the end of the book. We shall only outline some of the key reactions that lead to the formation of alcohol, although we will occasionally digress to describe the coenzymes or enzymes that are involved.

The Initial Phosphorylation of Glucose

If glucose and inorganic phosphate are added to dialyzed juice from yeast, no fermentation occurs and no sugar phosphate esters are formed, showing that one of the coenzymes must play a part in the reaction. If ATP is added to the dialyzed juice, however, phosphorylation of the sugar (i.e., the addition of phosphate to the sugar) is begun again, and hexose diphosphate can be isolated. Work with highly purified enzymes has clarified this series of reactions.

The first step in the series consists of the transfer of one of the phosphate groups of ATP to glucose to form glucose-6-phosphate and ADP, in a reaction that is catalyzed by an enzyme, *hexokinase*. All enzymes concerned with reactions of this type that involve ATP are called *kinases* (fructokinase, glucokinase, etc.). Next, the glucose-6-phosphate molecule, after a series of transformations, reacts with ATP again to form the hexose diphosphate, as is indicated in Fig. 10. The hexose diphosphate is now split into two triose phosphate molecules (three carbon sugars, each of which contains a phosphate). Although the initial products of the splitting reaction are different, they can be converted into one another in a reversible reaction (Fig. 11). The crucial intermediate in the formation of alcohol is glyceraldehyde-3-phosphate.

Fig. 10.   Initial steps in the metabolism of glucose-phosphorylation.

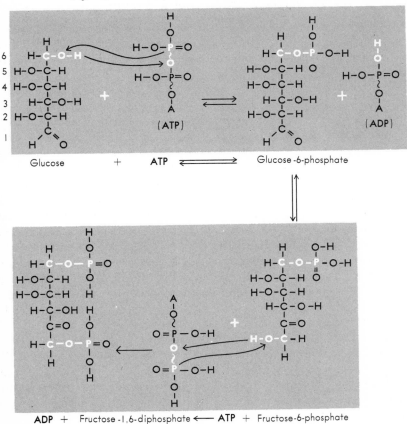

Glucose   +   **ATP** ⇌ Glucose-6-phosphate

ADP + Fructose-1,6-diphosphate ⟵ **ATP** + Fructose-6-phosphate

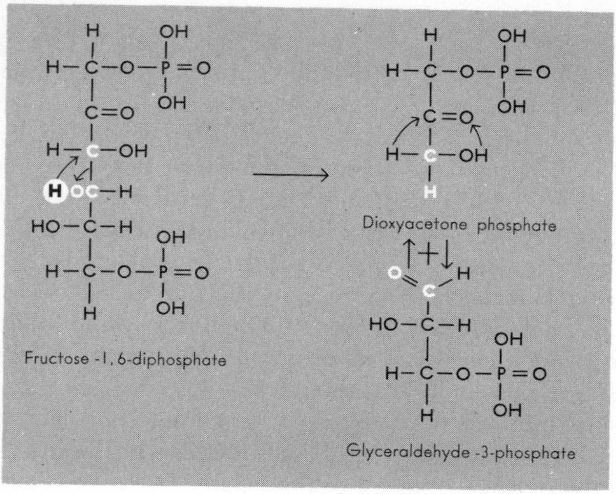

Fructose -1, 6-diphosphate

Dioxyacetone phosphate

Glyceraldehyde -3-phosphate

Fig. 11. Splitting of fructose-1, 6-diphosphate.

The Oxidation of Glyceraldehyde-3-Phosphate

Earlier we saw what happens to an aldehyde when it is oxidized to an acid. In the yeast juice there is an enzyme, triosephosphate dehydrogenase, which catalyzes this oxidative step, in which hydrogen is removed. The glyceraldehyde-3-phosphate adds to the SH group of the enzyme, as shown in Fig. 12, and then oxidation occurs to form the "energy-rich group" on the enzyme. This group, in turn, reacts with inorganic phosphate to form 1,3-diphosphoglyceric acid. In the next stage, the energy-rich phosphate group is transferred from carbon 1 to a molecule of ADP to form 3-phosphoglyceric acid and ATP, a reaction that is catalyzed by an enzyme called phosphoglyceric acid kinase.

The coenzyme of the triosephosphate dehydrogenase plays a vital role as an oxidant in the reaction. As we discussed previously, Harden and Young demonstrated that a small heat-stable cofactor is required for alcoholic fermentation. This dialyzable cozymase (coenzyme 1) was soon isolated and identified as a compound containing adenylic acid, nicotinic acid amide (a B vitamin effective in the prevention of pellagra), and phosphate (Fig. 13). These findings provided the first revealing clue to the function of vitamins in cellular metabolism. As we progress into various phases of biochemistry, we shall see what a significant part the vitamins perform in the coenzyme molecules.

After the structure of coenzyme 1 was established, it was referred to as diphosphopyridine nucleotide (DPN). The reactive group of DPN (and of TPN—a coenzyme which differs from DPN by having another phosphate attached to the ribose of the adenylic acid-triphosphopyridine nucleotide) in biological oxidations is the pyridine ring. This ring can accept two electrons and a proton to form a new structure. In the oxidation of glyceraldehyde-3-phosphate, two hydrogen atoms are removed, but in the reduction of DPN, only one hydrogen ion is added to the ring. Since two electrons are transferred to DPN, a hydrogen remains in the medium ($DPN^+ + 2H^+ + 2e \rightarrow DPNH + H^+$), and the reduced DPN is designated as DPNH. In the oxidation of glyceraldehyde-3-phosphate, therefore, the coenzyme (DPN) becomes reduced, 3-phosphoglyceric acid is formed, and ADP is converted into ATP. Once all the available DPN has been reduced, fermentation will cease, but as we shall see later,

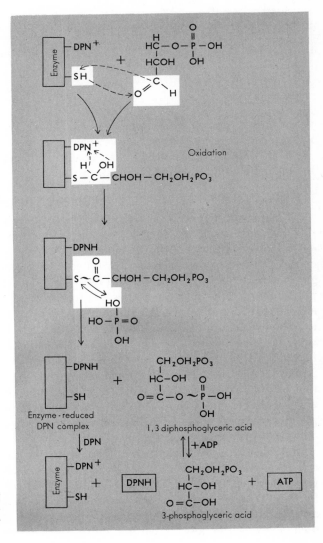

Fig. 12. Oxidation of glyceraldehyde-3-phosphate. Role of DPN (coenzyme I).

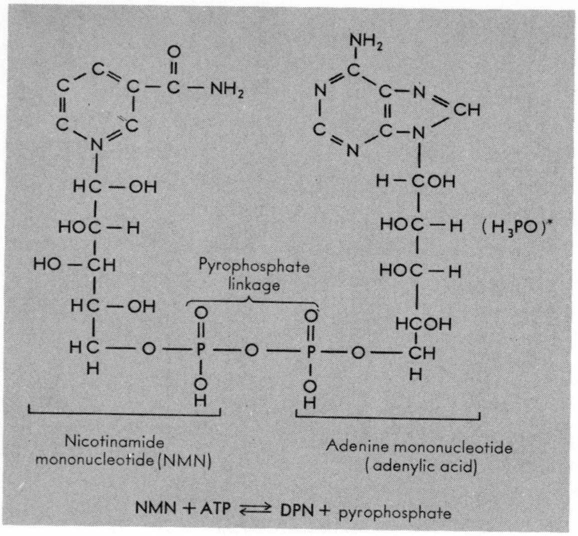

Fig. 13. Structure of DPN and TPN.

* The addition of a phosphate group on the 2 position of the ribose part of adenosine converts DPN into another coenzyme known as triphosphopyridine nucleotide (TPN). TPN also functions in electron transport.

yeast juice contains other enzymes which catalyze the oxidation of DPNH, thus regenerating the active coenzyme. Note that inorganic phosphate as well as ADP must be continually supplied in order to remove the energy-rich intermediate formed on the enzyme surface.

Metabolism of 3-Phosphoglyceric Acid

The yeast juice further metabolizes the 3-phosphoglyceric acid in the following manner. The 3-phosphate ester is converted into the 2-phosphate ester (2-phosphoglyceric acid), which is then dehydrated, causing the energy in the molecule to be redistributed into an energy-rich phosphate group, phosphoenolpyruvic acid (Fig. 14). The formation of phosphoenolpyruvic acid from 2-phosphoglyceric acid is catalyzed by an enzyme called *enolase*. In the presence of phosphate, enolase is strongly inhibited by fluoride, because of the removal of magnesium (as magnesium fluorophosphate) which is essential for the action of enolase. If fluoride is used in a yeast juice fermentation, then, the metabolism of 2-phosphoglyceric acid can be prevented, a fact that enabled biochemists to accumulate 3-phosphoglyceric acid in quantities that could be isolated and identified. They were then able to demonstrate that when 3-phosphoglyceric acid is added to an uninhibited yeast extract, it is rapidly converted into phosphoenolpyruvic acid.

This new energy-rich phosphate group is no different from the one formed in the oxidation of glyceraldehyde-3-phosphate. The phosphate in the presence of a specific kinase can be transferred to ADP to form ATP and pyruvic acid. Pyruvic acid is an example of an α-keto acid, and, as we mentioned previously, the first step in the metabolism of such acids is the loss of carbon dioxide (decarboxylation). In yeast extracts, the decarboxylation reaction leads to the production of carbon dioxide and acetaldehyde ($CH_3CO-COOH \rightarrow CH_3CHO+CO_2$). The enzyme that catalyzes this reaction is called carboxylase and requires as a cofactor (cocarboxylase) another one of the B vitamins, thiamine. For thiamine to be enzymatically functional, it must be phosphorylated to form thiamine pyrophosphate (TPP). In the final stage of the production of alcohol, acetaldehyde is reduced to alcohol by the hydrogens that were passed on to DPN in the oxidation of glyceraldehyde-3-phosphate. The reduction of the acetaldehyde to alcohol or the oxidation of alcohol to form acetaldehyde is catalyzed by an enzyme called alcohol dehydrogenase. The coenzyme necessary for the action of this enzyme is DPN ($CH_3CHO+DPNH+H^+ \rightleftarrows CH_3CH_2OH+DPN$).

A summary of these remarkable reactions that are catalyzed by yeast is shown in Fig. 15. The unraveling of each step, the isolation and purification of the numerous enzymes and coenzymes, and the establishment of the scheme for the whole yeast cell represent some of the most significant and magnificent developments in the history of biochemistry and cellular physiology. The over-all results of this series of reactions, starting with glucose, are: First, for each molecule of glucose fermented two molecules of alcohol and two of carbon dioxide are formed. Secondly, for each molecule of glyceraldehyde-3-phosphate oxidized, one molecule of DPN is reduced and later reoxidized at the expense of the molecule of acetaldehyde that is formed from the decarboxylation of pyruvic acid. Two triose molecules are produced from each glucose molecule, leading to the formation of two molecules of alcohol and CO_2.

Thirdly, two molecules of ATP are employed in the phosphorylation of each molecule of glucose. The second ATP is used in the formation of

Fig. 14. Metabolism of 3-phosphoglyceric acid.

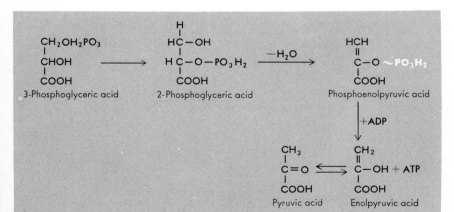

hexose diphosphate, each molecule of which yields two molecules of glyceraldehyde-3-phosphate; each of these takes up a molecule of inorganic phosphate after it has been oxidized to form two energy-rich phosphate groups that are transferred to ADP to produce ATP. Therefore, up to this stage in fermentation the yeast has recovered only the amount of ATP used in the initial stages. In the dehydration of the two molecules of 2-phosphoglyceric acid, however, two additional energy-rich phosphate groups are formed. In alcoholic fermentation there is thus a net synthesis of two molecules of ATP for each molecule of glucose metabolized.

Note that the splitting of pyruvic acid is probably the only irreversible reaction in the entire fermentation process. The conversion of fructose-1,6-diphosphate back to fructose-6-phosphate does not lead to the reformation of ATP. The "reversibility" of this reaction depends on the removal of the phosphate in the number 1 position by hydrolysis. The reaction is catalyzed by a specific enzyme (phosphatase). Note the importance of inorganic phosphate and ADP as controlling factors in cellular metabolism.

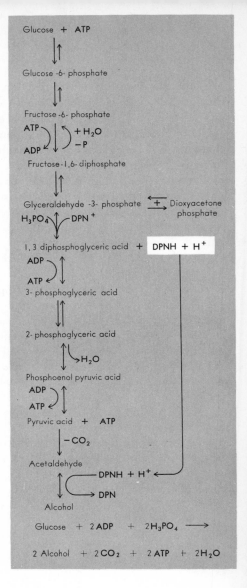

Fig. 15. Summary of the reactions in alcoholic fermentation.

6

Glycolysis

For a long time after the discovery of the phosphorylation of hexoses in alcoholic fermentation by Harden and Young, the process was not considered significant except as a means of shaping the hexose molecule for fermentative breakdown. Similar studies of other cells and particularly of muscle, however, revealed that phosphate plays a dominant role in energy transformation, especially in that involved in muscle contraction. The questions confronting the early investigators of muscle contraction were: What is the chemical source of the energy that powers the muscle machine, and how is this energy transformed into the mechanical energy of contraction?

The investigators soon uncovered these facts: (1) muscle can contract in a normal manner in the complete absence of oxygen; (2) lactic acid is produced during anaerobic contractions and accumulates with continued stimulation until the muscle becomes fatigued; (3) if the fatigued muscle is then put into oxygen, it recovers its ability to contract, and lactic acid simultaneously disappears; (4) less lactic acid is formed in a muscle that has access to oxygen than in one that contracts anaerobically. All the information pointed to the existence of a proportionality between the amounts of work done, of heat produced, of tension developed in a muscle, and the quantity of lactic acid formed.

By the late 1920's, it became evident that some of the energy expended in anaerobic muscle contraction comes from the conversion of glycogen (which is stored in muscle) to lactic acid in a process called *glycolysis*. For many years, physiologists and biochemists thought that the energy-supplying glycogen breakdown occurred just when the muscle contracted, but Hill found, after carefully measuring the heat during the contraction of a single muscle fiber, that there were two kinds of heat in addition to the initial heat associated with contraction: a relaxation heat and a delayed heat output called recovery heat. When he placed muscle under anaerobic conditions, however, he observed the contraction and relaxation heat but no recovery heat, and concluded that oxygen was needed to obtain the recovery heat. Since lactic acid accumulated in his experiment, there still appeared to be a close correlation between glycolysis and muscle contraction.

At about this time, a new chemical substance was isolated from muscle which in acid solutions readily broke down into creatine and phosphate. The compound turned out to be creatine phosphate and to be energy-rich just like the phosphate in ATP (see Fig. 16). Since it seemed to disappear during muscle contraction, it was thought to be related in some way to the energy supply and glycolysis. When the pure compound became available, large amounts of heat were found to be liberated when the compound was hydrolyzed, thus proving conclusively that a great quantity of potential energy was stored in the nitrogen-phosphorus bond. But one still had to admit that muscle contraction seemed to be closely connected with the breakdown of glycogen into lactic acid.

In 1930 the puzzle began to clear up when Lundsgaard demonstrated that muscle contraction could proceed anaerobically without lactic acid being formed, provided he poisoned the muscle with a compound called iodoacetic acid. He reported that this non-lactic acid contraction was accompanied by a pronounced breakdown of creatine phosphate, and when the latter was exhausted the muscle would no longer contract. From earlier work, the amount of heat produced when creatine phosphate is hydrolyzed was known. When researchers measured the quantity of creatine phosphate broken down during muscle contraction and compared it with the expected amount of heat, they found that an exact

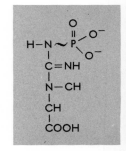

Fig. 16. Creatine phosphate.

relationship existed between the amount of creatine phosphate broken down and the amount of heat produced. In other words, when muscle contracts anaerobically, without glycogen breakdown, the heat production can be accounted for by the breakdown of creatine phosphate.

Investigation at this time into the biochemistry of the breakdown of glucose and glycogen by mammalian tissues, particularly by muscle cells, revealed that intermediates in the breakdown of these substances are essentially the same as those found in alcoholic fermentation. Lundsgaard was able to trigger muscle contractions without producing lactic acid, because the iodoacetic acid combined with the SH group in the triose phosphate dehydrogenase and prevented the oxidation of glyceraldehyde-3-phosphate. Creatine phosphate thus disappeared as a result of the inhibition of ATP synthesis. If dialyzed muscle extract were used, creatine phosphate was broken down only when ADP was present, indicating that creatine phosphate did not directly react with the muscle proteins. Subsequent work by Englehardt showed that the muscle protein (ATPase) reacted with ATP to break it down into ADP and inorganic phosphate. This was the first clear suggestion that ATP might be the immediate energy source for muscle contraction. Additional evidence demonstrated that creatine phosphate acts as a reservoir of high-energy phosphate which can re-phosphorylate ADP to form ATP in this reversible reaction:

$$CP + ADP \rightleftharpoons ATP + C$$

The reaction is catalyzed by an enzyme called creatine phosphokinase. ATP is thus generated in the glycolytic process just as it is in alcoholic fermentation, and when excess ATP is available, the high-energy phosphate is transferred to creatine. When a muscle that has been poisoned by iodoacetic acid contracts, the ATP can be generated only at the expense of this creatine phosphate reservoir.

Although much is known about the contractile proteins of muscle, the mechanism of converting the chemical energy of ATP into mechanical work remains obscure. In the chemical composition of muscle, there are two important contractile proteins, actin and myosin. When actin and myosin are mixed they form a complex (actomyosin), which can be made into "synthetic" threads that rapidly decompose ATP and at the same time are capable of contracting. We can be reasonably certain, then, that actin and myosin are the major proteins of the contractile muscle fiber. The mechanical arrangements of these fibers in the muscle and the function of ATP in causing contraction are gradually being clarified, particularly for skeletal muscle. The present view is that the contraction is produced by the sliding of actin and myosin fibers over one another. Since ATP under appropriate conditions is known to dissociate or to separate actin and myosin, when the nerve impulse stimulates a muscle to contract, the first event may be the activation of the ATPase system. Destruction of the ATP would allow re-association of the actin and myosin, resulting in a contraction, without changing the shape of the individual fibers.

The Initial Steps in Glycolysis

Glycogen is a complex polymer of glucose. As shown in Fig. 17, the sugar molecules are linked together by a bond (glycosidic) between the number 1 carbon of one glucose and the number 4 carbon of the next glucose. Branching from this straight chain, which may have as many as 18 glucose units, are a number of 1–6 glycosidic linkages that then continue as 1–4 linkages. Glycogen, therefore, is a branched, complex polysaccharide with a molecular weight as high as four million. Enzymes attack it at either the 1–4 or the 1–6 linkages. Many hydrolytic enzymes (glycosidases-amylases) can break down glycogen to smaller polysaccharides or free-glucose units. In muscle, when glycogen is broken down it reacts first with inorganic phosphate instead of water at the 1–4 glycosidic linkage; the process is thus a phosphorolysis, not a hydrolysis.

The ensyme that catalyzes this reaction is phosphorylase, and the

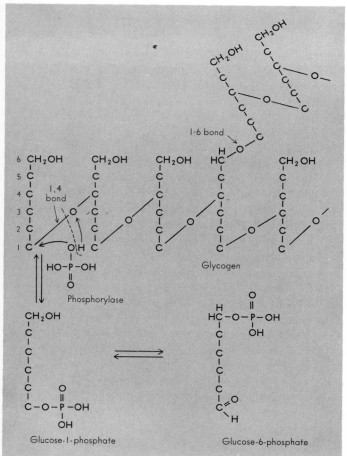

Fig. 17. Initial step in glycogen metabolism.

product of the reaction is glucose-1-phosphate and the remainder of the glycogen molecule. Once the outer tier of 1–4 glycosidic linkages is broken, the reaction stops. There are hydrolytic enzymes, however, that split the 1–6 linkage and thereby expose additional 1–4 glycosidic linkages which phosphorylase will attack. The new phosphate ester formed in the breakdown of glycogen is readily converted into glucose-6-phosphate by the enzyme, phosphoglucomutase; this phosphate ester, you will recall, is the compound formed when ATP reacts with glucose. All these reactions are reversible, and if there is an excess of glucose and ATP, glycogen synthesis will occur. Recent evidence indicates that this is probably not the primary pathway for the biosynthesis of glycogen. In principle, however, the reactions are the same.

After the formation of glucose-6-phosphate, glycolysis and alcoholic fermentation follow a common path until pyruvic acid is formed. Here the pathways again diverge, for muscle, unlike yeast, does not contain carboxylase, the enzyme that decarboxylates pyruvic acid. Pyruvic acid will, nevertheless, react with the reduced DPN and oxidize it in much the same way as the acetaldehyde did in the yeast extract. When pyruvic acid is reduced by DPNH, lactic acid is formed, and the enzyme that catalyzes this reversible oxidation-reduction reaction is called lactic dehydrogenase. The net effect of the anaerobic reaction sequence in muscle is that for every glucose unit (from glycogen), two molecules of lactic acid are produced. DPN is alternately reduced and oxidized, as it is in alcoholic fermentation.

In yeast juice, as in muscle extract, the sequence generates four new energy-rich phosphate bonds for each glucose molecule metabolized. In the fermentation of free glucose, however, two of the new bonds are used in the preliminary phosphorylation of the glucose molecule, so there is a net gain of only two molecules of ATP. In the first stage in glycolysis, on the other hand, the glycogen is split by inorganic phosphate and not by ATP, so that when the product, glucose-1-phosphate, is converted into glucose-6-phosphate, only one molecule of ATP is needed to produce each molecule of hexose diphosphate. Muscle glycolysis, therefore, gains a net of three molecules of ATP for each glucose unit of glycogen metabolized, enabling us to classify glycogen as an energy reservoir. Although the ester link of glucose phosphate that is formed in the uptake of free glucose uses ATP, the energy is not lost when the glycosidic linkage in glycogen is created, for this energy is such that inorganic phosphate can react with it to form again a hexosephosphate ester, which can be metabolized.

The control of glycogen or glucose metabolism in muscle is very similar to the process described for yeast. The reoxidation of DPNH, the action of inorganic phosphate, and the utilization of ATP are the essential processes. Can you describe now the chemical events that take place in muscle during a 440-yard dash? What happens to ATP? If the inorganic phosphate increases in concentration, what happens to glycogen breakdown?

In muscle glycolysis, then, the same fundamental mechanisms—oxidation-reduction, dehydration, and phosphorylation—are in operation as are present in alcoholic fermentation. These reactions are the basic ones involved in energy and carbon transformation in fermentations. In the past twenty-five years, we have found that in the metabolism of various organisms, these relatively simple steps account for almost all the fermentation products formed from carbohydrates by the organisms. For example, lactic acid bacteria produce lactic acid by a sequence of reactions identical to those observed in muscle. Outlined in Fig. 18 are a number of reactions that are carried out by different organisms. Note that in the formation of most of these compounds, only simple oxidation-reduction, hydrolytic, or decarboxylation reactions occur, and their net effect is to produce a number of electron acceptors that then oxidize the reduced DPN, which is formed in the triose phosphate dehydrogenase reaction. Pyruvic acid is formed in the same way by these organisms.

The Oxidation of Reduced DPN by Other Oxidants

Thus far we have discussed the oxidation reactions in which DPN acts as the electron acceptor. This step is the important initial dehy-

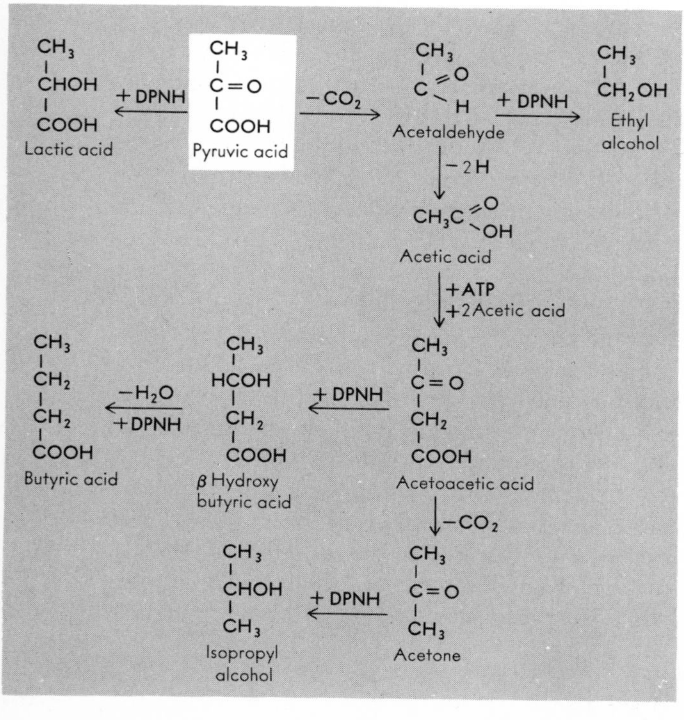

Fig. 18. Some products of fermentative metabolism.

drogenation in the metabolism of a number of substrates. In a series of brilliant studies beginning in the early 1920's, Warburg found traces of two other important coenzymes in cells grown under aerobic conditions. He was intrigued by the ability of iron compounds—blood charcoal in particular—to *catalyze* the oxidation of many different organic substances, using molecular oxygen as the electron acceptor. Pure charcoal made by heating sucrose does not possess this property, and Warburg attributed the catalytic action of blood charcoal to its iron content. From this and other evidence, he concluded that an iron-containing substance in the cell is essential for the activation and utilization of oxygen.

The importance of iron compounds in oxygen utilization was also suggested by the observation that low concentrations of cyanide and carbon monoxide inhibit respiration. Warburg found that cyanide blocked iron-catalyzed oxidations and concluded that it affected cellular respiration by combining with an iron-containing "respiratory enzyme" (Atmungsferment). In addition, he observed that the respiratory inhibition by carbon monoxide could be reversed by visible light. Since earlier experiments by others had shown that carbon monoxide would combine with hemoglobin but could be dissociated with visible light, Warburg concluded that his respiratory enzyme was an iron compound very much like hemoglobin.

Subsequent studies by Keilin clearly revealed the importance of a number of these heme pigments for electron transport. He called these cellular pigments *cytochromes* and was able to demonstrate at least three different heme proteins, which he named cytochromes a, b, and c respectively.

Warburg and Christian also isolated from cells a second coenzyme that was needed to set off oxidative reactions. They were able to isolate from yeast a yellow protein capable of catalyzing the transfer of electrons from reduced DPN to some other acceptor. They removed the yellow component by dialysis and prepared it in pure form. It proved to be a riboflavin compound that could accept electrons from DPNH. Since reduced flavins are colorless, the two investigators followed the reduction process by observing the disappearance of the yellow color. As might be expected, it is not the free riboflavin that is effective in electron transport, but the nucleotide derivative.

Two forms of riboflavin involved in electron transport occur in nature. One is flavin mononucleotide or riboflavin phosphate, and the other is flavin adenine dinucleotide (Fig. 19), which is analogous to DPN, a substance formed from another B vitamin, nicotinic acid amide. The flavins can exist in either the oxidized or the reduced form, and DPN is one of the better reducing agents in the presence of certain enzymes. Although reduced flavins are readily reoxidized by oxygen when they are free in solution, they are often poorly oxidized when bound to an enzyme. The reasons for this are not entirely clear.

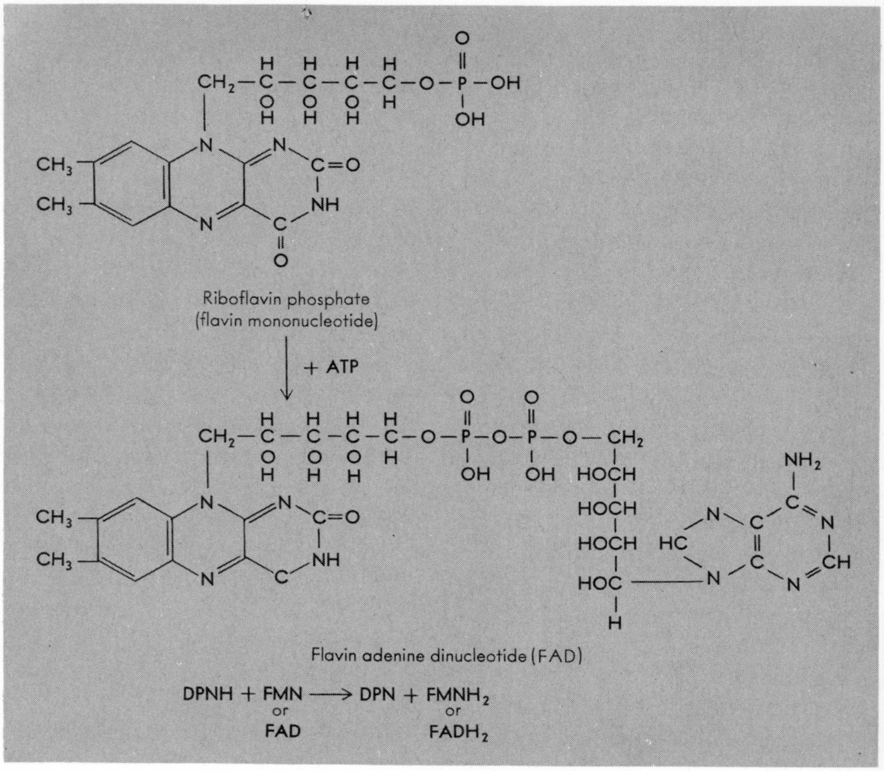

Fig. 19. Riboflavin coenzymes.

The flavin operates as an intermediate carrier of hydrogen between reduced DPN and other electron acceptors before oxygen is reduced. A number of outstanding investigations have indicated that the *cytochromes* are the oxidizing agents for the reduced flavins. Cytochrome pigments are present in all aerobic cells and can be alternately reduced and oxidized by a number of agents. Extensive studies have proved that the cytochromes contain a protein and an iron porphyrin group (Fig. 20). The iron porphyrin of the cytochromes is made up of four pyrrole rings and an atom of iron, as shown in Fig. 20, and is very similar to the pigment in hemoglobin. Cyanide and carbon monoxide are potent poisons of these iron-containing catalysts. In cytochromes, it is the iron in the porphyrin that undergoes oxidation-reduction: $Fe^{+++} + e \rightarrow Fe^{++}$. One of the cytochromes, cytochrome C, appears to be particularly important in the transfer of electrons to oxygens. The porphyrin of this cytochrome attaches itself to the protein by forming a linkage with two sulfur atoms in the cysteine residues of the polypeptide chain. An additional enzyme is necessary to oxidize the reduced cytochrome C by molecular oxygen and is called *cytochrome C oxidase*. The evidence indi-

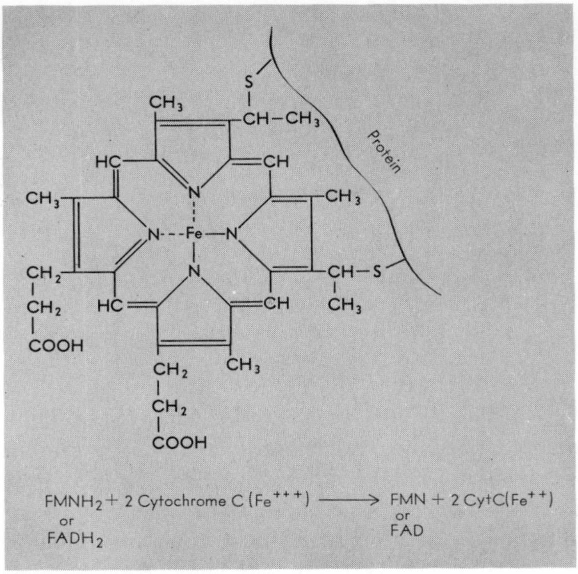

$$FMNH_2 + 2 \text{ Cytochrome C } (Fe^{+++}) \longrightarrow FMN + 2 \text{ CytC}(Fe^{++})$$
or   or
$$FADH_2 \qquad\qquad\qquad\qquad\qquad\qquad\qquad FAD$$

Fig. 20. Cytochrome C.

cates, therefore, that the mechanism of the electron transport to oxygen is from the flavin coenzymes to the cytochromes and then to molecular oxygen. In all aerobic organisms studied, the pathway of electron flow seemed to occur in the following sequence: DPN → flavins → cytochromes → molecular oxygen.

Carbon Dioxide and the Formation of Electron Acceptors

Succinic acid ($COOHCH_2CH_2COOH$), a four-carbon dicarboxylic acid, is manufactured in a number of organisms. For many years its formation puzzled biochemists, since it was not obvious how a substance containing four carbon atoms could arise from a six-carbon sugar or a three-carbon compound such as pyruvic acid. Earlier work indicated that the CO_2 tension in a medium containing growing cells tends to increase the productivity of succinic acid. While studying glycerol fermentation in propionic acid bacteria, Wood and Werkman first established that CO_2 uptake (fixation) in non-photosynthetic organisms is an essential process in the formation of key compounds that can accept electrons from DPNH.

Subsequent isotope work with C^{13} and C^{14} labeled CO_2 revealed that when carbon dioxide is in some way added to the methyl group of pyruvic acid, a four-carbon acid called oxaloacetic acid is formed. The oxaloacetic acid, when reduced with DPNH, forms malic acid. When it loses water, malic acid gives rise to fumaric acid which can be reduced by DPNH to produce succinic acid. The sequence of reactions is shown in Fig. 21. Thus CO_2, by combining with pyruvic acid, forms key compounds that can act as electron acceptors.

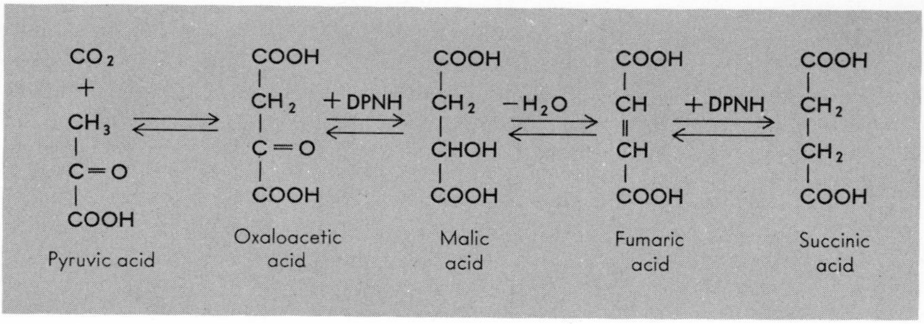

CO_2		$COOH$		$COOH$		$COOH$		$COOH$
$+$		CH_2 $+DPNH$		CH_2 $-H_2O$		CH $+DPNH$		CH_2
CH_3 $\rightleftharpoons$		$C=O$		$CHOH$		$\parallel$		CH_2
$C=O$		$COOH$		$COOH$		CH		$COOH$
$COOH$						$COOH$		
Pyruvic acid		Oxaloacetic acid		Malic acid		Fumaric acid		Succinic acid

Fig. 21. Formation of succinic acid.

When Szent-Györgyi in his early experiments found that compounds such as succinic and fumaric acid will stimulate respiration of tissues, he considered them to be significant in the hydrogen transport process and thus assigned them a catalytic role. The function of these four carbon dicarboxylic acids in electron transport remained a mystery, however, until Krebs discovered that they are involved in a cyclic process of electron transport. The catalytic effect of these acids on respiration, explained Krebs, results from their participation in a cyclic series of reactions that is needed for the oxidation of pyruvic acid to CO_2 and water. In these reactions the electrons are transported to oxygen over the DPN-flavin-cytochrome system. The net effect of this series of reactions is the complete combustion of pyruvic acid to CO_2 and water.

Considering the number of glycosyl units used, glycolysis is an inefficient mechanism for the synthesis of ATP, since a great deal of energy is still available in the fermentation products. As we indicated in the last chapter, the oxidative metabolism of pyruvic acid is the most effective method of generating phosphate-bond energy in aerobic organisms. This stage of metabolism of carbohydrates is called the Krebs citric acid cycle, and it stands at the intersection where the main routes of cellular metabolism converge. This cyclic metabolic machine handles products derived from the metabolism not only of glucose but also of amino acids and fatty acids.

Pyruvic Acid Oxidation

In one of the initial key reactions, the metabolism of pyruvic acid, carbon dioxide is removed and an "active" (energy-rich) two-carbon unit is formed. This initial reaction of pyruvic acid is very similar to the one that occurs in the formation of alcohol, since the first step is a decarboxylation of pyruvate to form an active aldehyde. The main difference here, however, is that the two-carbon fragment formed is not free acetaldehyde, but is combined to the coenzyme (thiamine pyrophosphate) of the decarboxylase (Fig. 22). This activated aldehyde is transferred to a second coenzyme, lipoic acid, which contains a disulfide (S—S) link. In this transfer, the S—S bond is reduced by the active aldehyde, leading to the formation of an SH group

7

Oxidative Metabolism

and the energy-rich acetyl group attached to the second sulfur atom.

The active two-carbon unit is very similar to the three-carbon energy-rich unit that forms in the triose phosphate dehydrogenase reaction. This active C_2 unit is now transferred to an SH group in another coenzyme (coenzyme A), resulting in the formation of completely reduced lipoic acid (2SH) and acetyl coenzyme A. The reduced lipoic acid reacts with DPN to form reduced DPN and the original active form (S—S) of the lipoic acid molecule. The net result of this series of rather complicated reactions (outlined in Fig. 22) is the formation of an energy-rich two-carbon unit, DPNH and CO_2. Note that the energy liberated in this *oxidative decarboxylation* is not lost as heat, as is the case when free acetic acid is formed as the product. If acetyl CoA is hydrolyzed to form CoA and acetic acid, large amounts of heat are given off.

The structures of the coenzymes involved in pyruvic acid metabolism are shown in Fig. 23. Note that sulfur is an important element in the structure of all three and that the B vitamin, pantothenic acid, is part of the molecule of coenzyme A. The isolation and identification of coenzyme A was one of the major advances of modern biochemistry. Not only is coenzyme A of intrinsic interest as an active biochemical reagent, but it is essential for many diverse reactions. Although the exact mechanisms of the series of reactions outlined in Fig. 22 are not clearly understood at the present time, the formation of the main product (acetyl-CoA) is well established. Acetyl CoA is the key product of the metabolism of a number of organic acids. Additional metabolism that takes place, through

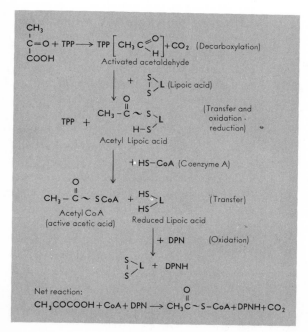

Fig. 22.   Oxidative decarboxylation of pyruvic acid.

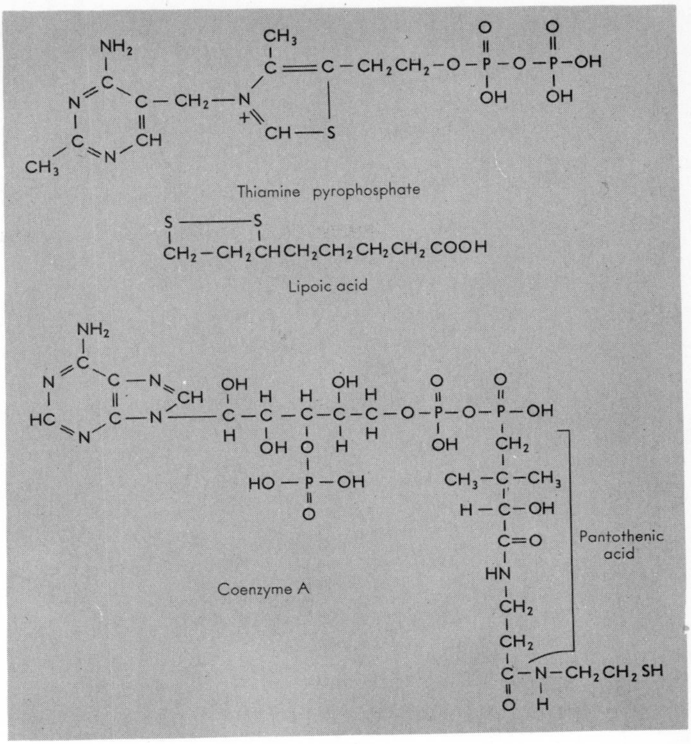

Fig. 23. Coenzymes for pyruvic acid metabolism.

oxidation of this C_2 unit in the citric acid cycle, is linked to reactions that transfer electrons ultimately to molecular oxygen. Consequently, in the presence of oxygen the reactions go to completion and liberate a large amount of energy that is conserved as adenosine triphosphate.

The citric acid cycle is initiated by the condensation of the active acetic acid with oxaloacetic acid to form citric acid, a six-carbon compound containing three carboxyl groups (Fig. 24). Note that the energy of the acetyl—CoA link is used in the condensation, and, as a result, the coenzyme A molecule is readily removed in the presence of water. In the ensuing reactions, citric acid is rearranged, by the removal and addition of water, to form isocitric acid.

As shown in Fig. 25, isocitric acid is then oxidized to form a new keto acid which rapidly loses CO_2 to form α-ketoglutaric acid. Since this latter keto acid is metabolized in a series of steps that are analogous to the initial steps in pyruvic acid metabolism, the oxidative decarboxylation of α-ketoglutaric acid results in the formation of succinic acid and CO_2. Succinic acid is then oxidized to fumaric acid, which, upon hydration, leads to the formation of malic acid. The oxidation of malic acid regenerates the oxaloacetic acid and thus closes the cycle by entering the initial reaction, the condensation of acetyl CoA to form citric acid. When all the necessary enzymes are present, this cycle can be initiated by the addition of any of the compounds within the cycle. To keep the cycle going, however, acetyl CoA must be supplied.

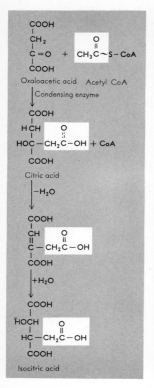

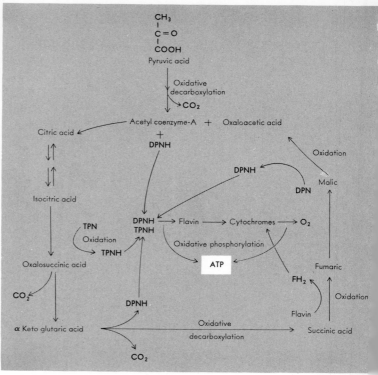

Fig. 24. (left) Synthesis of citric and isocitric acids.

Fig. 25. (right) Citric acid cycle.

This integrated group of enzymes—the dehydrogenases, the decarboxylases, and so forth—which engineer the complex series of reactions of the Krebs cycle are localized in the mitochondria of the cell. Bound to this structure in some unknown way are the various coenzymes, which include coenzyme A, DPN, TPN, the flavin coenzymes, as well as the cytochromes. The mitochondria from several different kinds of cells have been isolated, and have been shown to be capable of carrying on the final oxidative stages of cell metabolism by themselves. Of considerable significance is the fact that the reduced coenzymes (DPNH or TPNH) can be oxidized by the mitochondria with the liberation of energy. This oxidation, which is known to require the flavins as well as the cytochromes, is one of the crucial events in oxidative metabolism. The energy liberated in the oxidation of these cofactors is used to synthesize adenosine triphosphate.

As we saw in Fig. 25, ATP is formed in three different stages of electron transfer. The electron flow from the high-energy level of reduced DPN to the lower energy level of the flavin coenzymes thus liberates sufficient energy to form one ATP molecule. The same is true for the reduction of the cytochromes by the reduced flavins and the reduction of oxygen by the reduced cytochromes. The oxidation of one mole of DPNH by the mitochondrial enzyme complex therefore leads to the synthesis of three moles of ATP. Consequently, the transfer of two electrons (2H) to oxygen gives rise to the esterification of three inorganic phosphate molecules. Since one oxygen atom is used ($2H + O \rightarrow H_2O$), the ratio of the phosphate taken up to oxygen used is $3(P/O = 3)$. Note that in the oxidation of one pyruvic acid molecule, four molecules of reduced pyridine nucleotide are formed, and the oxidation of them leads to the synthesis of twelve ATP molecules. Two ATP molecules are obtained in the oxidation of succinic acid and one in the oxidation of ketoglutaric acid (not shown in the figure). The total combustion of pyruvic acid to CO_2 and H_2O thus leads to the net synthesis of fifteen ATP molecules. It is no wonder that the mitochondria have been called the powerhouses of the cell. *Oxidative phosphorylation* is one of the most interesting and intriguing problems of modern biochemistry. Unfortunately, we know very little about the mechanisms involved in the generation of ATP in oxidative processes of this type.

Although the citric acid cycle is most often considered in terms of the oxidation of a substrate and the generation of ATP, it is also valuable in the production and utilization of the carbon skeletons of many other compounds. For many years, the breakdown of carbohydrates was thought to be primarily a degradation process whose sole purpose was the liberation of energy. This is true as far as energy production is concerned, but we now know that the formation of key intermediates in the breakdown of carbohydrates is very significant in the synthesis of compounds of biological interest. Indeed, it has been proposed that during the rapid growth of the cell, the principal function of the citric acid cycle is to supply the cellular carbon skeletons for biosynthesis.

It is also significant that during evolution various cell types have not developed different structures or systems for the metabolism of sugar fermentation products and other cellular foods. The mitochondrion, for example, also functions as the cellular furnace for the combustion of fatty acids, amino acids, and other fuels.

Fatty Acid Oxidation

The oxidation of fatty acids proceeds in several distinct steps, but the final product is the active acetic acid unit, acetyl-CoA. Fatty acids, then, feed two carbon units into the citric acid cycle in the same way that pyruvic acid does. One crucial oxidative step in fatty acid metabolism we must now consider. The fatty acids of biological importance include a large series of straight-chain acids that begins with formic acid and continues through acetic acid to compounds with more than twenty carbon atoms. Interestingly enough, the naturally occurring fatty acids are predominantly even-numbered.

Let us examine the metabolism of the four-carbon butyric acid. The first thing discovered about fatty acid oxidation was the fact that the mitochondria contain the necessary enzymes and cofactors for this process. It was next observed that a small amount of ATP is essential to initiate fatty acid oxidation, and subsequent developments have demonstrated that this activation by ATP is required for the formation of the coenzyme A derivative of fatty acids. The carboxyl group of the fatty acids, such as butyric, reacts with ATP to form an energy-rich adenylic acid derivative. This intermediate then reacts with coenzyme A to form butyryl-CoA.

As indicated in Fig. 26, the coenzyme A derivative is oxidized between the number 2 and 3 carbons to form a double bond. This is called β-*oxidation*. The electron acceptor in this oxidation is usually a flavin. Water is now added across this double bond to form an OH group on the beta carbon, a process that is analogous to the hydration of fumaric acid to form malic acid. Oxidation now occurs, leading to the formation of a double-bond oxygen on carbon number 3 that is capable of reacting with coenzyme A to split off two coenzyme A units. At present, the cycling of CoA in the breakdown of fatty acids through β-oxidation is probably a reasonable explanation for the metabolism of fatty acids of any chain length. Thus, if we start with an even-numbered fatty acid, we will obtain an even number of acetyl CoA units. A C_8 fatty acid, therefore, give rise sequentially to: C_8CoA → C_6CoA → C_4CoA → $2C_2$CoA. In each step, a C_2CoA is generated.

The metabolism of odd-numbered fatty acids presents some special problems. If we start with a C_9 fatty acid, the steps appear to be the same (C_9 → C_7 → C_5 → C_3) until we get to C_3CoA (propionyl-CoA). Present evidence indicates that this active form of propionic acid must take up CO_2 to form the even-carbon succinic acid before it can be further metabolized. This CO_2 fixation is rather complicated and not yet well understood. Although active C_2 units (acetyl-CoA) are used in the synthesis of fatty acids, the pathway is apparently not a simple reversal of the scheme shown in Fig. 26.

Note that in the oxidation of

Fig. 26. Fatty acid metabolism.

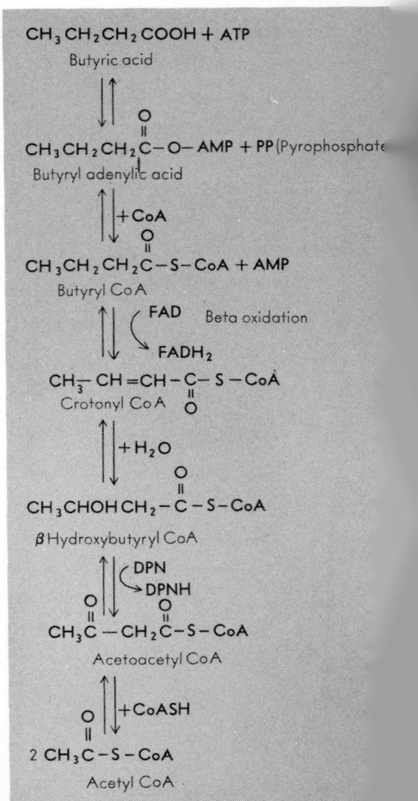

butyric acid we obtain two acetyl-CoA units, which, on entering the citric acid cycle, lead to the generation of thirty ATP molecules. In addition, five other ATP molecules are generated during the initial oxidative reactions that give rise to the formation of acetyl-CoA. It is evident, therefore, that much more energy can be obtained from the oxidation of fatty acids than from the equivalent carbon-length of a carbohydrate.

During starvation or in an extreme case of diabetes, ketones (acetone, acetoacetic acid, and β-hydroxybutyric acid) accumulate in the blood. The appearance of these bodies seems to be due to the hydrolysis in the liver of acetoacetyl CoA. Acetoacetic acid can be reduced to β-hydroxybutyric acid or can be decarboxylated to acetone. Although we know that the production of these acetone bodies is intimately linked to carbohydrate metabolism, the exact relationship is not entirely clear.

Amino Acid Metabolism

The proteins occupy a central place in both the structural and the dynamic aspects of living matter. Along with the nucleic acids, they compose the most important macromolecular structures of cells. Proteins are composed of amino acids, and it is partly the difference in amino acid composition that gives the proteins some of their unique properties. A supply of amino acids, either from the diet or from the biosynthetic machine, is obviously essential to cells and tissues for the synthesis of proteins. When these essential amino acids are acquired in the diet, the remainder of the nitrogen needed for protein synthesis can be supplied in the form of ammonium salts. Ammonia is an intermediate in nitrogen metabolism, and most organisms, when given an adequate amount of utilizable carbon compounds and other essential growth elements, can readily employ ammonia as their principal source of protein nitrogen. A crucial reaction in the uptake and incorporation of ammonia into proteins involves one of the intermediates in the citric acid cycle. As outlined in Fig. 27, α-ketoglutaric acid can be converted into the amino acid, glutamic acid, by a process called *reductive amination*, in which TPNH is the reducing power and ammonia is the nitrogen source. This reaction is reversible, and, in fact, the enzyme that catalyzes the reaction is called glutamic dehydrogenase. It is clear that this reaction

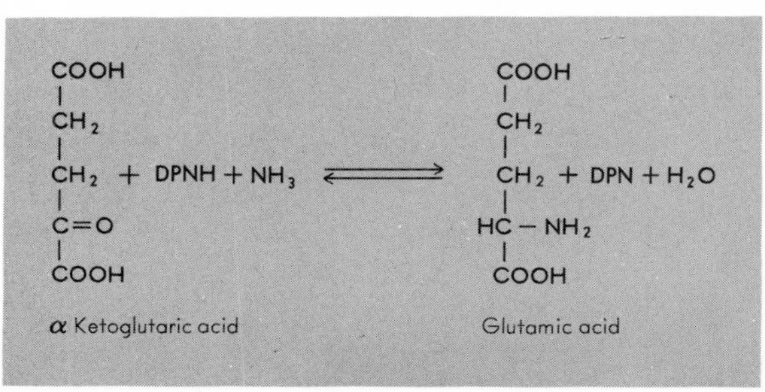

Fig. 27. Synthesis of glutamic acid.

$$\begin{array}{c} COOH \\ | \\ CH_2 \\ | \\ CH_2 + DPNH + NH_3 \\ | \\ C=O \\ | \\ COOH \end{array} \rightleftharpoons \begin{array}{c} COOH \\ | \\ CH_2 \\ | \\ CH_2 + DPN + H_2O \\ | \\ HC - NH_2 \\ | \\ COOH \end{array}$$

α Ketoglutaric acid　　　　　Glutamic acid

serves as a main link between amino acid and carbohydrate metabolism.

Once ammonia is converted into amino nitrogen, it can be transferred to other carbon skeletons to form a different amino acid. This process of *transamination*, shown in Fig. 28, involves an amino acid and a keto acid. Since the reaction is reversible, the process represents one of the principal metabolic pathways for the formation and deamination of amino acids. Enzymes called transaminases, which catalyze these various reactions, are known to occur in a variety of animal tissues, plants, and microorganisms. The B vitamin, pyridoxine, in the form of pyridoxal phosphate, is an essential cofactor for all transaminases. Two of the intermediates in the citric acid cycle, therefore, serve as carbon skeletons for the synthesis of two amino acids. The process of reductive amination and transamination can lead, as far as we know, to the incorporation of ammonia into a large number of carbon skeletons that are formed from carbohydrate and fatty acid metabolism. How particular carbon skeletons are formed, then, is the major problem in amino acid biosynthesis, and, for the most part, it appears that the intermediates of the citric acid cycle are the initial compounds for the formation of amino nitrogen.

The relationship between the citric acid cycle and amino acid metabolism is clearly shown in the series of reactions that lead to the synthesis of the amino acid, arginine, and to the major nitrogenous excretion product of man, urea. When excess amino acids are fed to animals, much of the excess nitrogen is excreted as urea. The unraveling of the mechanisms involved in urea synthesis was another milestone in our understanding of cyclic processes in biochemistry. As indicated below, the "ornithine cycle" regulates the removal of ammonium ions and depends for its functioning on the presence of the citric acid cycle. The essential compounds that are required to maintain the continued synthesis

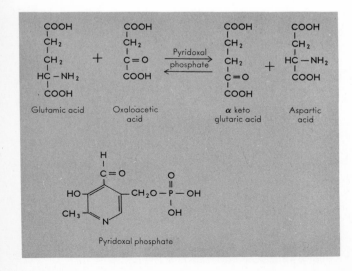

Fig. 28. Transamination.

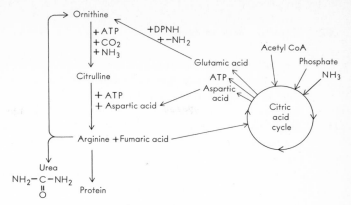

Fig. 29. Urea and arginine synthesis.

of urea are the following ones: ATP, aspartic acid, ammonia, and CO_2.

As shown in Fig. 29, glutamic acid can be converted into ornithine by reduction and transamination. In the presence of CO_2, ATP, and ammonia, ornithine is converted into citrulline by the addition of NH_2COOH, which is made from CO_2 and NH_3 in the presence of ATP and the appropriate enzyme. Aspartic acid adds a $-NH_2$ group to citrulline to form the amino acid, arginine. In the liver, if arginine is not used for protein synthesis, it is broken down by the enzyme, arginase, to form urea and ornithine. The ornithine or urea cycle thus leads to the excretion of two nitrogen atoms in the form of urea, and, as is evident from the scheme, the process depends on the immediate participation of the citric acid cycle for a supply of ATP and of the appropriate carbon-skeleton intermediates for the synthesis of aspartic and glutamic acids.

The mechanism of synthesis and degradation of a number of the amino acids is a fascinating subject, but it is not necessary for us to go into the details of these reactions at this time. The principles outlined for the synthesis and breakdown of glutamic acid, aspartic acid, and arginine hold for the metabolism of other amino acids. To pursue the many other interesting metabolic products of amino acid metabolism, some of which are extremely important to cellular function, you should consult a more advanced book on biochemistry.

The glycolytic or fermentative pathway of glucose metabolism represents the major route for the formation of pyruvic acid from carbohydrates. There are a number of other alternate pathways of carbohydrate metabolism, however, the most prominent of which has been called the oxidative or pentose phosphate shunt. This pathway has the advantage of providing a means for the combustion of glucose to carbon dioxide without the participation of the citric acid cycle. In addition, the pathway leads to the formation of ribose, the five-carbon sugar we have shown to be important in the synthesis of a number of coenzymes, ATP, and other cellular components. As discussed in another book in this series,[1] this pathway of carbohydrate metabolism is also intimately

---

[1] A.W. Galston, *The Life of the Green Plant* (Englewood Cliffs, N.J.: Prentice-Hall, 1961).

concerned in the carbon dioxide uptake in the photosynthetic organisms.

The first step in the metabolism of glucose by the oxidative pathway that is different from any of the steps in the fermentative pathway involves glucose-6-phosphate. As shown in Fig. 30, glucose-6-phosphate is oxidized to the corresponding 6-phosphogluconic acid. The electron acceptor in this case is TPN instead of DPN. The TPNH may be of considerable importance in the synthesis of a number of compounds of biological significance (fatty acids, etc.). If TPNH is not used as a reducing compound in biological synthesis, it can be oxidized by the flavin-cytochrome systems of the cell. Under these circumstances, molecular oxygen is the eventual electron acceptor.

Following the initial oxidation, 6-phosphogluconic acid is oxidized by TPN to form an intermediate which (see Fig. 30) is presumed to be 6-phospho-3-ketogluconic acid and is quickly decarboxylated to give the known products, CO_2 and ribulose-5-phosphate. Ribulose-5-phosphate is readily converted into ribose-5-phosphate by the enzyme, isomerase.

In the photosynthetic process, the primary acceptor of CO_2 is apparently ribulose-1,5-diphosphate. The latter is formed from ribulose-5-phosphate and ATP. Details of the various carbon skeleton transformations that occur in photosynthesis are beyond the scope of this study.

Summary

From the last three chapters, we can see that great advances in the analysis of energy transformation and biosynthetic processes in organisms have been made during the past thirty years. The fundamental energy-liberating reactions that are necessary to maintain the diverse and sometimes complex cellular machines are superficially relatively simple. These reactions are examples of electron or hydrogen transport. To trigger biochemical reactions, we simply couple energy-yielding reactions to energy-consuming steps. The special form of chemical energy employed by biological systems for these coupling reactions is that stored in the pyrophosphate bond of ATP.

As Krebs has recently emphasized, the first major stage of energy transformation in living matter terminates in the synthesis of ATP, at the expense of the free energy liberated during metabolism. In these energy transformations, we observe only a few basic processes, namely, oxidation, reduction, dehydration, hydration, decarboxylation, acetylation, phosphorylation, amination, and transamination. For carbohydrate, protein, and fat metabolism, the three key substances that are formed prior to any significant energy liberation are: acetyl CoA, α-ketoglutaric acid, and oxaloacetic acid. These are broken down in the citric acid cycle to liberate over two-thirds of the cellular energy. The reactions of various food materials are summarized in Fig. 31.

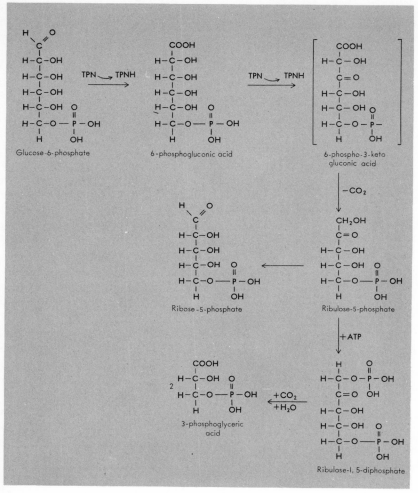

Fig. 30. Hexose monophosphate oxidation.

Fig. 31. Summary of the metabolism of fats, carbohydrates, and proteins (after Krebs).

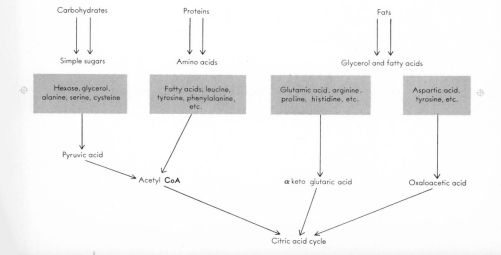

By using sunlight as a source of energy, green plants convert, by the process known as *photosynthesis,* carbon dioxide and water into all the organic molecules associated with life; the gas, oxygen, also appears as one of the products. At the same time, the green plants replenish the organic world, which is continuously being destroyed by heterotrophic organisms. Life on this planet, therefore, is absolutely dependent on light.

Photosynthesis is not the only biological process affected by light. The vision of all animals is sparked by light energy striking specialized cells in their eyes. The movement of plants and certain animals toward or away from light and many other rhythmic biological phenomena demonstrate that light energy is of overwhelming importance in the functioning of organisms. Before discussing these photobiological processes, we should consider the properties of light.

What is Light?

When a fine beam of sunlight is passed through a glass prism, it is resolved into its component colors: red, orange, yellow, green, blue, and violet. Newton demonstrated that if the light passes through a second prism that is reversed, the colored lights recombine to give white light; but if a single color is selected from the spectrum, no subsequent treatment can change it in any way. From this simple experiment, he concluded that white light is composed

8

Special Problems in Energy Transformation

64

of many colors that can be separated into discrete units or particles. However, when evidence was presented which indicated that light rays could bend and spread, much as water waves fan out when a rock is dropped in a pond of water, this corpuscular or particle theory was rejected in favor of a wave theory. Since light always diffracts, so that the edge of a shadow is never perfectly sharp, the wave theory is consistent with many of the classical phenomena related to optics.

Numerous experiments, nevertheless, have revealed that light also resembles ordinary material particles. These "light" particles, which are designated as photons or light quanta, were discovered in early experiments concerned with the effect of light on solid matter. When light was beamed onto certain metal plates, electrons were ejected. The velocity of the ejected electrons was uninfluenced by the intensity of the light, although the higher the light intensity the greater were the number of electrons ejected. This is the basic principle underlying the photoelectric cell. When light hits a metal surface, the ejected electrons generate a small electrical current. Researchers directing colored light (i.e., light with a specific wavelength) onto the plate found that the velocity of an ejected electron was affected in such a way that the kinetic energy of the electron was inversely proportional to the wavelength or color.

The photoelectric effect led Einstein to propose that light is in the form of small photons or quanta of energy. The energy of the photons determines the color of the light, and the entire energy of one photon is absorbed by an ejected electron, resulting in the destruction of the photon; the energy to eject the electron thus comes from that contained in the destroyed photon. Earlier, when Max Planck attempted to explain the energy distribution (in different wavelengths) of the light emitted by hot bodies, he discovered a fundamental constant that relates frequency to energy. This constant, h, is also the one needed to give the proper energies in the photoelectric effect. The energy, E, of a photon or light quanta could be calculated from the equation, $E = hv$, where v is the frequency and h is Planck's constant. Frequency has its usual meaning, normally expressed as the number of vibrations per second.

Consider the waves on the surface of a pond. When a rock is dropped into the water, a wave spreads out from the point of impact. If we drop rocks repeatedly, multiple waves are set up; the faster we drop the rocks, the closer are the peaks of the waves together. This distance between the peaks of two waves is the wavelength. Since the velocity of propagation of the wave remains constant, it is clear that the shorter the wavelength, the greater is the frequency. In other words, the speed of propagation divided by the wavelength equals the frequency. In the case of light, therefore, we can modify the equation describing the energy

of a photon to this expression, $E = \dfrac{hc}{\lambda}$, where c is the velocity of light,

(3×10^{10} centimeters per second), and λ is the wavelength of light in centimeters. The actual value of h is 6.624×10^{-27} erg-seconds.

In biochemistry, we commonly talk in terms of calories per mole of a substance instead of in terms of ergs, and we use a variety of units for the wavelength of light. Radio waves are customarily measured in meters or centimeters, and visible light in either millimicrons or Angstrom units. A meter $= 10^2$ centimeters $= 10^3$ millimeters $= 10^6$ microns $= 10^9$ millimicrons. The Angstrom unit is 10^{-8} centimeter and therefore equals 1/10 millimicron. By multiplying by the appropriate constants, we can write the above equation relating energy to the wavelength of light as:

$$E = \frac{2.86 \times 10^7 \text{ calories per mole}}{\text{wavelength in millimicrons}}$$

You will recall that a mole is equal to 6×10^{23} molecules of a substance (in a volume of a liter), and the photochemical equivalent would be, therefore, 6×10^{23} photons. Thus, by making our energy calculations on this basis, we can speak of calories per mole as we do in ordinary chemical reactions. Blue light, with a wavelength of 450 millimicrons, then, is equivalent to approximately 64 kilocalories/mole. We now see that different colors of light represent photons or quanta of different energies, those of blue light having much greater energy than those of red. We also note that the connecting link between the corpuscular and the wave theory of light lies in the fact that the energy of photons is directly proportional to the frequency of the light waves.

The word "light" is usually applied to what can be seen by the eye. Since scientists, however, speak of ultraviolet "light" and report that they can "see" other types of radiation by means of photocells and other cells, light to them means more than just the radiation that is visible to the unaided eye. The distinction between various types of light radiation is thus one of convenience; all radiations of the electromagnetic spectrum represent a single phenomenon, and their differences are only ones of wavelength and photon energy.

Molecular Structure and the Absorption of Light

We know that when light quanta fall upon a metal plate, electrons are ejected. Light quanta also eject electrons from molecules in solution. In an atom or a molecule in a stable state, there are as many electrons around the positive nucleus or nuclei as the atomic number of the constituents. These electrons occupy the space surrounding the nuclei; some are close in and tightly bound by electrostatic forces and some are farther away and therefore more loosely bound. The momentum and spin of these electrons fix them in orbitals about the nucleus. Only a certain number of electrons can fit into each orbit, and the ones that are most weakly bound are in the outermost filled orbit. According to quantum mechanics, other orbits are "allowed" even farther out, but since it requires energy to move an electron away from its positive nucleus, the electron will tend to stay as close to the nucleus as it can. All chemistry occurs in these outermost or valence orbits.

In a molecule, we usually assign two electrons to each of the outer orbits, since a neutral molecule usually has two electrons in its outermost filled orbit. The spins of these electrons must be in opposite directions, so we say that the electrons are "paired." (See Fig. 32.) If one of these electrons absorbs a light quantum, it can be boosted into a higher orbit, and the molecule is then said to be in an excited state.

The energy required to push an electron from one orbit out to

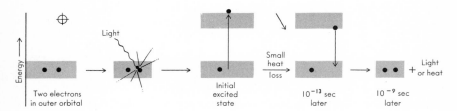

Fig. 32.   Absorption of light quanta by molecules.

Fig. 33.   The absorption spectrum.

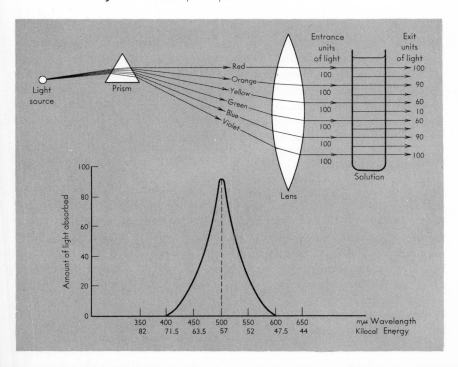

another obviously depends on the energy difference between the two orbits, and we can determine this difference by shining light of various wavelengths onto the molecules. Since the energy in a photon of light can only be used on an "all or none" basis (that is, a *part* of a quantum cannot be utilized), only those photons that are effective in forcing the electrons to the outer orbits will be absorbed, and by invetsigating which wavelengths of light are absorbed, we obtain what is called the *absorption spectrum* (Fig. 33). If we break up ordinary light through a prism, then beam the different colors through a solution of some chemical, and observe the light coming out the other side, we can determine qualitatively which wavelengths of light are absorbed by the molecules in solution. By using a photocell instead of our eye, we can quantitatively record the amount of light absorbed and plot this as shown in Fig. 33. Since the absorption peak is at 500 mμ, this means that approximately 57 kilocalories of energy per mole, on the average, are necessary to push an electron from a filled to an unfilled orbit.

Photosynthesis

Light is absorbed by the chlorophyll pigments in plants and transformed by the cells into the chemical energy needed for the synthesis of carbohydrates. Priestley was the first to observe that green plants are capable of producing oxygen, although he did not know that light is essential in manufacturing it. During the next few years, it was clearly established that green plants could also remove the CO_2 produced during metabolism. Some time later the quantitative relationship between the CO_2 taken up and the oxygen produced was established. Plants, like animals, also respire, and we now know that they remove hydrogen from carbon skeletons and transfer them to oxygen, and also break carbon chains to form CO_2. These processes occur in the dark as well as in the light. Photosynthesis, therefore, must be the reverse of respiration; the light energy must be used to transfer hydrogen from water back to a carbon skeleton, which eventually comes from CO_2. Thus, in its simplest form, photosynthesis may be written as follows:

$$CO_2 + 2H_2O + light \longrightarrow (CH_2O) + O_2 + H_2O$$

To make a complete sugar molecule, six CO_2 molecules must be available. Therefore:

$$6CO_2 + 12H_2O + light \longrightarrow C_6H_{12}O_6 + 6O_2 + 6H_2O$$

If it seems peculiar that we put in water in the two equations and then take out water, the explanation lies in the fact that the oxygen comes from the water and not from CO_2. In the photochemical process, water is presumed to split into a reducing part (H) and an oxidizing part (OH). The reaction of OH units is believed to give rise to some water and O_2.

The primary process in photosynthesis, therefore, is the absorption of light quanta by the chlorophyll molecule. Since chlorophyll absorbs in the red region of the spectrum (656 mμ)—for this reason the leaves look green—we should suspect that the red quanta are the effective ones in photosynthesis. In general, this has been found to be true. Although we are not sure of the exact mechanism, it is clear that light energy must cause some of the electrons of the chlorophyll molecules to jump to an excited state. This excitation energy somehow splits water to form a highly reducing "H" and an oxidant "OH." Thus, in the light phase of photosynthesis, the photochemical decomposition of water produces a tremendous amount of potential energy.

The active hydrogen does not appear as free hydrogen, however. Instead, it must be bound either to the chlorophyll or to some other cofactor. Eventually oxidation takes place and, interestingly enough, the first compound to be reduced is TPN. In this photochemical reduction of TPN, sufficient energy is available to generate an energy-rich phosphate group. The initial reducing agent formed in photosynthesis, whatever it might be, is quite energy rich, since it is capable of reducing TPN and, at the same time, of generating an energy-rich phosphate group. The production of ATP in the light reaction is called *photosynthetic phosphorylation*. As a result of this reaction, it is possible for green plants to produce ATP without oxidizing stored carbohydrate. Green plants are thus capable of producing an excess of carbohydrate (starch) which can be used by other organisms to supply their energy requirements.

One of the primary roles of TPNH is to reduce CO_2. Before this can occur, CO_2 must be taken up by some reaction in the plant. Recent isotopic evidence indicates that the CO_2 adds to a 5-carbon sugar, which immediately splits into 2 molecules of 3-phosphoglyceric acid. By reversing the glycolysis scheme and making use of reduced pyridine nucleotides and ATP, sugar can be synthesized (Fig. 34). If we determine the necessary energy balance for this series of reactions, which involves the fixa-

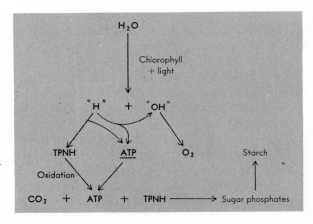

Fig. 34. Scheme of photosynthesis.

tion and reduction of CO_2, we conclude that 2.5 TPNH molecules are needed for the incorporation of 1 CO_2 molecule into glucose and the formation of 1 molecule of O_2. Since apparently only 1 H is formed per quantum of light absorbed, at least 5 quanta of light are necessary to form this amount of TPNH. Unfortunately, the experimental observations of various workers do not agree, since values ranging from 4 to 12 quanta have been reported, and much remains to be done to unravel the total problem of photosynthesis.

Vision

The visual process is another striking biological example in which excited states must be involved. In spite of a great deal of outstanding work on the structure of the retina and on the biochemistry of the visual pigments in the eye, very little is known about the basic mechanisms underlying vision. By exposing the optic nerve that connects the eye to the brain, researchers have demonstrated that a nerve impulse is produced by light shining on the retina, which indicates that the primary photochemical act must occur in the eye. A man's eye contains about 4 million units called cones and an additional 125 million units called rods; the rods lead to about a million optic-nerve fibers and are responsible for low-intensity vision in the dark, which is predominantly black and white vision. The cones control high-intensity color vision.

We assume that the triggering mechanism for the nerve impulse must lie in the rods and cones, because most of the visual pigments are found in the outer segments of them. One of the visual pigments (rhodopsin) has been extracted and has been shown to consist of a protein (opsin) attached to a derivative of vitamin A (retinene). Vitamin A is converted into retinene by the reduction of the alcohol group to the aldehyde by DPNH. Alcohol dehydrogenase catalyzes this reaction. When light strikes the visual pigments, rhodopsin immediately dissociates into opsin and retinene, and it is during this time that the nerve impulse is triggered. Oddly enough, the retinene that comes off is different from that which goes back on to opsin to make rhodopsin.

The relationship of these retinene isomers to vitamin A is shown in Fig. 35. The connection between vitamin A and vision was first noticed when a vitamin A deficiency was observed to affect the eyes' ability to adapt to the dark. When insufficient vitamin A is present, the reconstruction of the visual pigments is retarded and night blindness results. The absorption spectrum of rhodopsin is identical to the sensitivity of the eye to different wavelengths of light. The peak absorption and sensitivity is around 500 mμ.

How the photo-excitation of pigment molecules produces an impulse in the fibers is not understood. Early experiments suggested that at least the initiating step was clear. Rhodopsin is known to dissociate when illuminated with visible light and to reform in the dark. The obvious

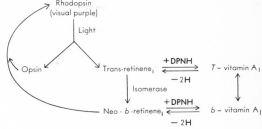

Vitamin A₁

Fig. 35. Some chemical events in vision.

conclusion was drawn from these facts: the photocurrent is associated with the changes in retinene. Recent studies, however, have indicated that these changes probably do not constitute the primary event, since the photochemical changes of rhodopsin are quite slow compared to the rate of initiation and conduction of the nerve impulse. It is now believed that the small chemical changes that occur during the visual process are due to some alteration that precedes the transformation in retinene itself. Maybe the photocurrent is caused by some electrolytic process that may result from the SH groups on opsin being exposed when retinene dissociates.

Bioluminescence

As we have indicated, when a quantum of light is absorbed by an atom or a molecule, an electron is boosted to a higher energy level. When the displaced electron returns to the original ground state, it can emit a quantum of light (the mechanism is called *photoluminescence*).

There are two fundamental types of photoluminescence: *fluorescence* and *phosphorescence*. Fluorescence is the emission of light from substances only during the time they are exposed to radiation of various kinds, or for a very short time afterwards, while phosphorescence persists after the exciting radiation is cut off. In many cases, phosphorescence is at a longer wavelength (lower energy) than fluorescence, and this tells us something about the mechanism. As indicated in Fig 36, in fluorescence the electron returns immediately to the ground state after excitation—by immediate, we mean in about 10^{-9} second. Usually this fluorescent light is at a longer wavelength than the exciting light, since some energy is lost as heat before the electron returns to the ground state.

In some cases, the electron may reverse its spin and move to an even lower energy state, but if it does it is almost impossible for it to return to the ground state until the spin is again reversed to the original direction, since no two electrons with the same spin can occupy the same orbit. The result is a long-term emission, a phosphorescence, of relatively low intensity that continues until something happens to reverse

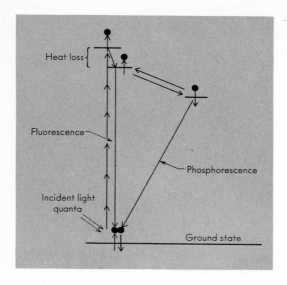

Fig. 36. Excitation of molecules to fluorescence or phosphorescence.

the spin. This emission may last for several seconds. We have described the fluorescent and phosphorescent states in some detail because these properties of molecules are becoming more and more important in the analysis of biological processes. When riboflavin phosphate is irradiated with ultraviolet light, for example, it emits a very brilliant fluorescence with a peak in the yellow-green at 530 mμ. When the flavin is reduced or when it combines with a protein, this fluorescence disappears or is greatly reduced, revealing valuable information about such reactions.

Although photoluminescence is the most common type of luminescence, other types are attracting the interest of scientists. *Electroluminescence*, for example, results when electrical energy excites molecules or atoms, and this type of cold light will probably provide much of the home lighting of the future. *Chemiluminescence*, a special concern of the biologist, is light emission caused by a chemical reaction.

The emission of light by organisms, *bioluminescence*, is a particular example of chemiluminescence. This form of luminescence occurs when the energy released during a biological oxidation forces an electron of a molecule to a higher energy level. In this case, the oxidation energy serves the same purpose as the incident proton does in photoluminescence. That organic chemicals could give off light was first observed in the latter part of the nineteenth century. In 1887 the French physiologist, Dubois, suggested that the light from a luminous clam, *Pholas dactylus,* is caused by an oxidizable substance he called *luciferin.* Luciferin remains stable when heated, but in the presence of oxygen and an enzyme, *luciferase,* it is destroyed by an oxidative reaction which gives out light. Scientists pursuing this problem have discovered that luciferin is not a single compound common to all luminous organisms. Apparently, there are many different luciferins, just as there are many different vitamins.

Much work remains to be done, therefore, on the various luminous forms, which range from the bacteria that give off a blue light (495 mμ) to the South American railroad worm (larva of a beetle), which gives off a red light (640 mμ). The light of decaying wood is produced by fungi (530 mμ). The luminescence of the sea is caused by a variety of forms: protozoa, radiolarians, sponges, jellyfish, comb jellies, brittle stars, snails, clams, squid, shrimp, copepods, fish, and many others. Probably the best-known luminous forms on land are the fireflies and glow worms. In addition to these luminous beetles, there are luminous spring-tails, flies, centipedes, millipedes, earthworms, and snails. The only known fresh-water luminous animal is the limpit (Latia), a native of New Zealand.

The mechanism of bioluminescence is poorly understood. We are confronted with two fundamental, unanswered questions concerning this type of luminescence: (1) what is the nature of the excited molecule? and (2) what is the chemical reaction that is capable of supplying so much energy for the excitation process? As we have seen, in most oxidations of biological interest, the energy is liberated in small units, usually just sufficient to synthesize a pyrophosphate bond (8 kilocalories/mole). Since *visible* radiation is emitted by organisms, much more energy must obviously be liberated by this oxidation process. Although a great deal of progress has been made in recent years in probing the nature of the chemical substances required for light emission, the basic mechanism remains obscure. In some respects, bioluminescence is very similar to photosynthesis and vision. In the latter two cases, the excited states are generated by light and chemistry results, while in luminescence the excited state is created by chemical reactions and the energy is lost as light.

A number of investigators have attempted to determine the importance of light emission to those organisms that display this remarkable ability. Unfortunately there is no clear answer as yet. There are many examples in which the bioluminescent reaction has been adapted to good biological use. The reproduction cycles of a number of organisms in the sea are intimately tied to light emission. The flash of the firefly is used by some species as a sex signal. Light emission by deep-sea organisms provides the only light source for those organisms at great depths that have eyes. In some cases luminous bacteria are known to grow in special glands on fish and to provide a regular source of light. Whether light emission in the deeps of the ocean trigger other photobiological processes is not known, but it is a safe guess that it does. A photograph of luminous toadstools is shown in Fig. 37.

The fact that there are many different chemical types of bioluminescent reactions suggests that they arose independently during evolution and are an expression of the reverse process of photosynthesis. In photosynthesis oxygen is evolved and organic molecules are produced by various reduction reactions, while in bioluminescence oxygen decom-

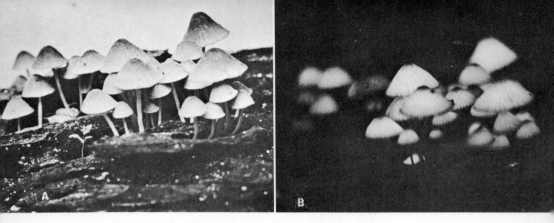

Fig. 37. Luminous toadstools. (Courtesy Dr. Y. Haneda.)

poses organic molecules in an oxidative event that leads to an excited state and light emission. It has been suggested that excited states of this type normally occur in oxidative reactions, but the energy is most often used to generate ATP rather than to produce light emission. In fact, one can present a reasonable argument that life arose and evolved by making use of excited states, as plants are capable of doing now in photosynthesis. During the early evolution of life, therefore, many more light-emitting organisms may have existed than we see at the present time. Bioluminescence is perhaps gradually being lost as organisms evolve effective means of utilizing the energy of the excited state. The use of light emission for specific purposes may thus have been secondary in the evolutionary process. Obviously, much work remains to be done in this area of biological energy transformation.

The structure, function, and metabolic activity of a cell or tissue depend on the presence of specific protein molecules, for proteins play an essential role in all phases of the chemical and physical activities associated with life. The synthesis and function of enzymes (one class of proteins) determine the nature of cellular products. Enzymes are also intimately associated with such physiological processes as muscular contraction, nerve conduction, excretion, and absorption. Since the specific antibodies capable of combining with foreign or disease-producing agents are proteins, as are most of the important hormones that regulate cellular and tissue function, to analyze the physiology and biochemistry of cellular activity in some detail we must know as much as possible about these important macromolecules. We shall discuss first the methods of isolation and purification of proteins and then consider their structure. In the next chapter, we shall investigate the nature of enzyme catalysis and, in the last chapter, the control of protein synthesis and cellular activity.

Proteins are composed of amino acids bound together by an amide linkage, called the peptide bond (see Chapter 1). One of the important properties of protein, its electric charge, depends on a fundamental property of the amino acids. Amino acids behave both as weak acids and weak bases, since they each contain at least one carboxyl group (—COOH) and one amino group (—NH$_2$). Such compounds are called

9

Proteins

75

ampholytes. An amino acid such as glycine, therefore, can carry both a positive and/or a negative charge, depending on the pH of the solution (Fig. 38). The addition of hydrogen ions (H^+) to a solution of glycine suppresses the ionization of the carboxyl group, and the molecule acquires a net positive charge. On the other hand, adding a base (OH^-) removes a proton from the ammonium group, resulting in a net negative charge. At a certain pH, the molecule is electrically neutral (the number of $+$ charges equals the number of $-$ charges), and this pH, at which the dipolar ion will not migrate either to the positive or negative pole in an electric field, is called the *isoelectric point.* This point can be determined by titration with acid and base as outlined in Chapter 3 or by electrophoresis as discussed below. The pH of the isoelectric point depends on the dissociation constants of the basic and acidic groups.

Proteins behave as ampholytes because in the formation of a peptide bond the ampholytic character of the amino acids is preserved. A tripeptide is shown in Fig. 39. Note that on the right side we have a free acid group and on the left a free amino group; R^1, R^2, and R^3 refer to the remainder of the carbon skeletons of the three amino acids. The number of acidic and basic groups in proteins depends on the number and types of amino acids present. Obviously, if we linked a hundred glycine molecules together by peptide bond formation, we would end up with a large polypeptide with one free carboxyl and one free amino group.

There are amino acids, however, which have more than one basic or acidic group. Aspartic acid, for example, has two carboxyl groups, and, consequently, even when it is in the middle of a peptide where the terminal or α carboxyl is part of the peptide bond, an acidic group (the β carboxyl) is still free. Another acidic amino acid is glutamic acid.

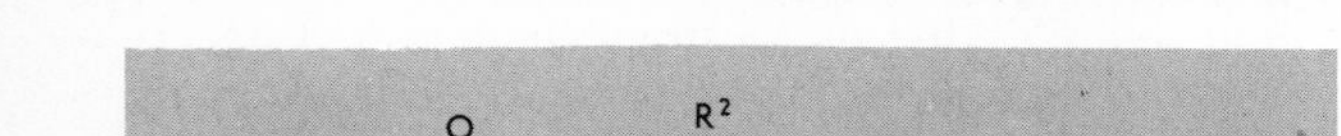

Fig. 38. Amphoteric properties of amino acids.

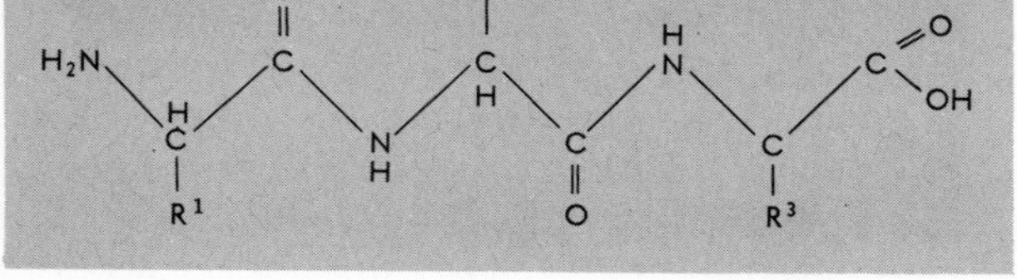

Fig. 39. A tripeptide.

Some amino acids, however, carry extra basic groups. Arginine, for example, has the following structure:

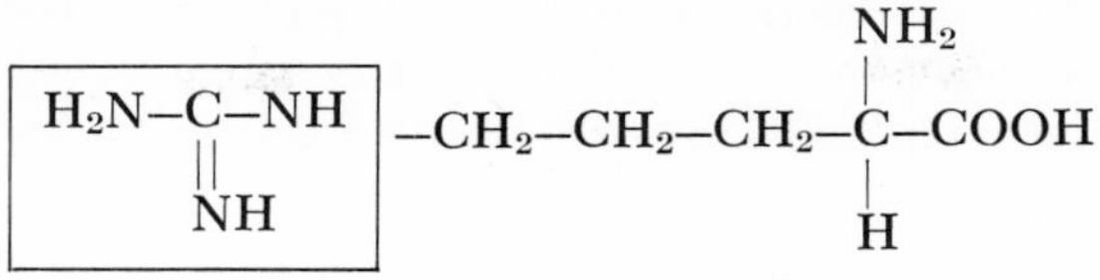

$$
\boxed{
\begin{array}{c}
\mathrm{H_2N-C-NH} \\
\parallel \\
\mathrm{NH}
\end{array}
}
\quad
\mathrm{-CH_2-CH_2-CH_2-\overset{\overset{\displaystyle NH_2}{\mid}}{\underset{\underset{\displaystyle H}{\mid}}{C}}-COOH}
$$

The guanido group shown in the square on the left is strongly basic. Another basic amino acid is lysine.

The behavior of proteins thus stems largely from the amino acid composition and the pH of the environment. All that has been said about the charge properties of amino acids can be repeated for proteins, since they, too, have an isoelectric point and will migrate in an electric field. The pH relative to the isoelectric point determines whether they move to the positive or negative pole. The isoelectric point of proteins depends on the relative numbers of free carboxyl and free amino groups, which in turn depend on the amino acid composition.

Isolation of Proteins

To study the specific properties of enzymatic activities of a particular protein, we must first isolate it from other cellular components. Although often a difficult task, those who have patiently tried one procedure and then another and finally obtained a pure, homogeneous product have had the satisfaction of being the first to see a unique biochemical event. Certain general rules should be observed in isolating a particular protein. We need to know (1) something about the stability of the protein at different temperatures and pH and (2) in what solvents, other than water, the protein can be precipitated without destroying its properties. With these points in mind, we shall now discuss briefly the various procedures used to purify proteins.

EXTRACTION FROM THE CELL

We usually start by making a crude extract from cells or tissues that contain the desired protein, and this is most often done at a low temperature (2–5°C). The cells are broken open by a number of methods: by grinding, freezing, and thawing; by shaking with glass beads; by disintegrating them with sonic waves, etc. Water or a buffer is normally used; proteins are often associated with other cellular structures, such as lipids, however, and may not be readily extracted with aqueous solvents. In some cases, the complex can be broken by washing the tissue first with cold acetone or butyl alcohol.

ISOELECTRIC PRECIPITATION

Since the solubility of proteins varies with pH, by carefully varying the pH of the solution we can often precipitate the desired protein without destroying it—or we can at least eliminate other unwanted proteins. Even though the purification may not be perfect, this procedure is a useful one for concentrating the protein and at the same time eliminating a large number of compounds with low molecular weights. The precipitated protein can then be dissolved in a small amount of buffer.

SALT FRACTIONATION

High concentrations of salt are often effective in precipitating proteins from an aqueous solution. Some proteins are very soluble in high salt concentrations and others are readily precipitated, thus enabling a number of proteins to be separated. Ammonium sulfate has been the most useful compound for this salting-out purpose, and pH has also been an effective variable in this process. We employ salts and pH as variables for isolating proteins because they are multiply charged, and such charged groups can react with the water molecules or with each other. If proteins tend to react with water molecules, their solubility is ordinarily increased. When ammonium sulfate reacts with the water, however, its effective concentration is lowered, and eventually the protein molecules will react with each other to form a precipitate (i.e., they become insoluble in water).

DIALYSIS

Some proteins (Euglobulins) are insoluble in pure water because the protein-protein interaction is stronger than the protein-water interaction. Such proteins can be brought into solution by the addition of a small amount of salt which interacts with the charge groups on the protein. The Euglobulins, then, can be separated from other proteins by dialysis against water that lowers the salt content of the protein solution.

SOLVENT FRACTIONATION

Organic solvents such as acetone and ethanol are miscible with water and are therefore effective in the precipitation of proteins. Such solvent fractionations are usually carried out at temperatures below the freezing point of pure water.

In general, the methods discussed above are the most useful ones for the purification of reasonably large amounts of proteins; they are always able to keep the protein concentration at a high level, which is an important advantage, since proteins in dilute aqueous solution tend to unfold and denature. Additional methods have also been applied, and we will discuss them below when we consider purity.

Criteria of Purity

SOLUBILITY

Since a homogeneous protein has a definite solubility in a given solution at constant pH, these factors enable us to establish an accurate criterion of purity. If we add increasing amounts of protein to a solution, determine the amount that is dissolved, and graph the result, we should find that the curve for a pure protein breaks sharply, as shown in curve A of Fig. 40. If the preparation contains two proteins, there should be two breaks in the curve. This technique for determining homogeneity is difficult to perform; in recent years other methods, depending primarily on electrical charge and molecular weight, are used more frequently.

ELECTROPHORESIS

Proteins, as we have seen, migrate in an electric field except at the isoelectric point, which is the pH where the net positive charge equals the net negative charge. Since a protein's net charge varies with pH, two proteins with different isoelectric points will obviously have different mobilities and can be separated by the technique of electrophoresis. This technique is most effective when the protein solutions are buffered and at a known pH. For most proteins (whose isoelectric points are below 7), the pH is made alkaline (7.5–8.0) to give them a maximum net negative charge. The solution is then placed in the bottom of a U-tube, and a buffer solution of the same pH is carefully placed on top of it (Fig. 41A). Electrodes are placed in the two arms of the tube, and the electric current is turned on. The negatively charged protein will move upward in the arm containing the positive electrode and downward in the arm containing the negative electrode (Fig. 41B).

We measure the rate of movement of the protein in the electric field by observing, through very sensitive optical instruments, the change in protein concentration at the boundary between the protein and the buffer (Fig. 41C). The change affects the optical properties of the area between the buffer and the protein solution; the incident light beam is bent the most as it passes through the solution in the region of the boundary, i.e., where the *protein gradient* is greatest. By using certain lenses, we can project a *schlieren band* (Fig. 41E), the peak of which occurs at the boundary. The area under this curve is an expression of the concentration of protein. As shown in Fig. 41F, only one such peak appears with a pure

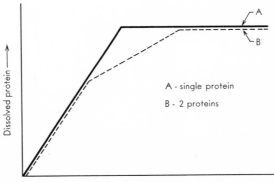

Fig. 40. Solubility curve of proteins.

protein. If two proteins are present, however, a discontinuity exists at the moving boundary, and two distinct patterns are formed (Fig. 41G). If these two proteins are similar in electric charge, just a shoulder may appear (Fig. 41H). A change in pH may change the mobilities of these two components, and better separation (and thus a better test for purity) can be achieved if the electrophoresis is carried out at a number of dif-

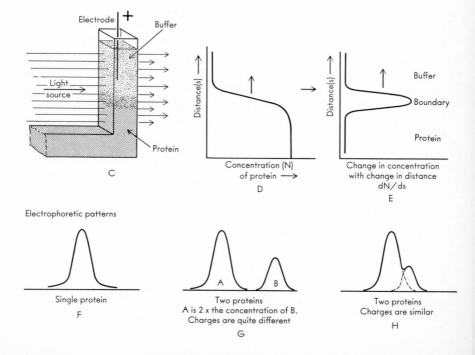

Fig. 41. Electrophoresis.

Buffer

Protein

A   B

Analysis of buffer-protein ascending (+) boundary

Electrode +   Buffer

Light
source

Protein

C

Distance(s) ⟶

Concentration (N)
of protein ⟶

D

Distance(s) ⟶

Buffer

Boundary

Protein

Change in concentration
with change in distance
dN/ds

E

Electrophoretic patterns

Single protein

F

A   B

Two proteins
A is 2 x the concentration of B.
Charges are quite different

G

Two proteins
Charges are similar

H

ferent pH values. Electrophoretic homogeneity, however, is not proof that the preparation is composed of a single molecular species. Molecules of different molecular weight may have similar migration rates in an electric field.

A modification of this electrophoretic technique has proved even better in isolating proteins. A protein is placed on a strip of agar, starch gel, sponge, or a piece of filter paper, which serves as a bridge between two tanks containing a buffer (Fig. 42). The movement of the protein in the electric field is slowed by the protein-absorptive capacity of the supporting material. The advantage of this method is the relative ease with which protein components can be observed. With this sensitive tech-

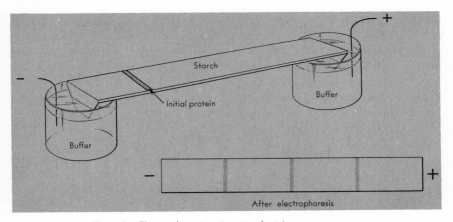

Fig. 42. Electrophoresis using starch strips.

nique, the heterogeneity of a number of highly purified preparations of proteins has been demonstrated. For example, nearly pure crystalline beef heart lactic dehydrogenase can be separated into three different proteins (Fig. 43). Since each of these proteins has enzymatic activity, we can fairly safely conclude that very small differences can exist in proteins that are otherwise identical in physiological functions. We will discuss this problem in greater detail when we take up protein synthesis.

Fig. 43. The separation of crystalline beef heart lactic dehydrogenase into three different enzymatically active components by starch gel electrophoresis. (Courtesy of Dr. C. L. Markert.)

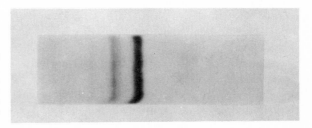

SEDIMENTATION

The large size of protein molecules allows us to observe their movement in a centrifugal field. The rate of their movement, other things being equal, is directly proportional to their size or molecular weight. In the centrifuge cell, the protein molecules move outward from the center of rotation, and a sharp boundary is formed between the pure solvent and the protein solution. By using the same optical method as in electrophoresis, we can study this boundary carefully. If one protein is present, we observe the same pattern as that at the bottom of Fig. 40). If there are two proteins, the heavier molecules will move ahead of the lighter ones. We must emphasize that apparent homogeneity in the experiment does not prove that the protein preparation is pure; under other conditions or by other procedures, heterogeneity may be observed.

As suggested above, the rate of sedimentation is an index to the molecular weight of macromolecules. Shape, however, as well as size influences the sedimentation rate, but for the purpose of calculating the molecular weight, the diffusion constant of the protein molecules adequately takes into consideration the influence of shape. With these two constants, we can calculate the molecular weight.

Protein Structure

Proteins, as we know, are made up of a number of amino acids linked together in a definite sequence by a peptide bond. Some proteins may contain more than one peptide, and these are held together by specific cross links (i.e., disulfide bonds). The arrangement or sequence of the amino acids in the polypeptide is called the *primary structure* of the protein. In most proteins, the tightly coiled polypeptide chain produces a helical shape which we call the *secondary structure* of the protein molecule. Many of the interesting biological properties of proteins result from this secondary structure, and we will consider it after discussing the amino acid composition.

AMINO ACID COMPOSITION

If proteins are treated with relatively strong acids (6N HCl) at 100°C for 10–20 hours, all the peptide bonds usually break and the amino acids are liberated. In recent years, a number of methods have been developed for determining, qualitatively and quantitatively, the amino acid composition of such a mixture.

A method called *paper chromatography* is often used in the qualitative determination of the amino acid composition. The process is based on the fact that amino acids will separate on a strip of paper according to the relative solubilities of the acids in two different solvents. A small drop of an aqueous solution of an amino acid mixture is placed on a strip of paper which is placed in a closed box or cylinder so that the bottom of it dips into an organic solvent (alcohol, etc.) saturated with

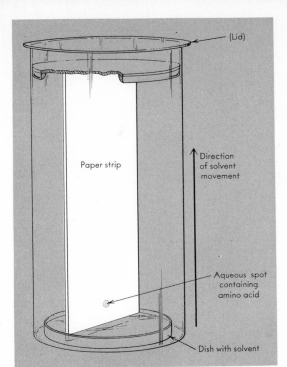

(Lid)

Paper strip

Direction
of solvent
movement

Aqueous spot
containing
amino acid

Dish with solvent

Fig. 44. Ascending paper chromatography.

water (Fig. 44). The organic solvent moves into and up the paper, passing over the water spot containing the amino acids. Those amino acids that are highly soluble in the organic medium compared to their solubility in water will be carried along with the solvent, while those that are relatively more soluble in water will remain behind. Under constant conditions, each amino acid moves a specific distance up the paper. After the solvent has climbed a suitable way, the paper is removed, dried, and sprayed with ninhydrin, a chemical that reacts with each amino acid to produce a readily observable colored spot for each one. Since each amino acid moves a specific distance, by comparing a given spot with known amino acids, we can often make preliminary identifications.

When a large number of amino acids are present, it is often convenient to chromatograph first in one direction on the edge of a large sheet of paper, using organic solvent, and then, after drying the sheet, to rotate it 90° and re-chromatograph, using a different organic solvent. The separation of a mixture of amino acids by this two-dimensional paper chromatography is shown in Fig. 45.

The principle of paper or partition chromatography has suggested other means of separating amino acids. For example, such substances as starch or silica gel can be substituted for the paper as a support for the stationary aqueous phase. The starch is combined with the organic solvent and water mixture and then packed in a column such as that shown in Fig. 46. The organic solvent, saturated with water, is added at the top of the column and allowed to flow until equilibrium between the

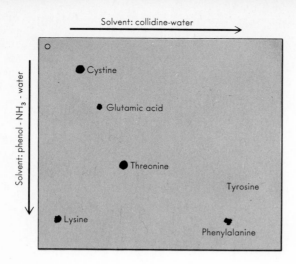

Solvent: collidine-water

Solvent: phenol - NH₃ - water

Cystine

Glutamic acid

Threonine

Tyrosine

Lysine

Phenylalanine

Fig. 45. Two-dimensional paper chromatography of amino acids.

water and organic phase is reached. A sample of the protein hydrolysate is then introduced at the top, and more solvent is allowed to pass through the column. The amino acids will move along the column at a rate that depends on their partition between the aqueous and the organic solvent phase. In addition, adsorption of the individual amino acids on the starch is different and thus aids in their separation. The differences in the movements of the amino acids are often large enough to enable the acids to emerge from the bottom of the column completely separated from each other. By collecting small samples of the effluent solution, we can often make a complete quantitative separation of all the amino acids. In Fig. 46, four individual amino acids are separated on such a column.

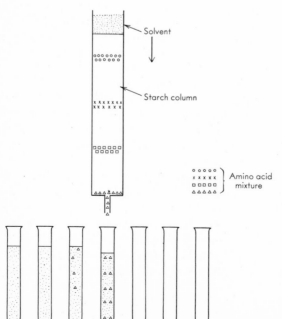

Solvent

Starch column

Amino acid mixture

Fig. 46. Chromatographic separation of amino acids on a column.

The best procedure for separating amino acids quantitatively is one that depends more on the chemical properties of the amino acids. Since amino acids have ionizable groups with different pK values, it is possible to separate them on columns which react with these charged groups. Ion exchange resins have proved extremely effective in this process and two general types, the cation exchangers and the anion exchangers, are used. When an amino acid mixture is added to such a column, the amino acids will exchange with one of the groups on the resin. The degree of exchange and the strength of the binding depend on a number of factors, including pH and the strength of the buffer. By continuously changing the buffer strength (gradient elution) and pH, we can quantitatively separate all the amino acids in a protein hydrolysate. With the aid of ingenious devices, we can continuously analyze the effluent from the column for its amino acid content. The results of a typical separation of a mixture of amino acids on a cation exchanger are shown in Fig. 47.

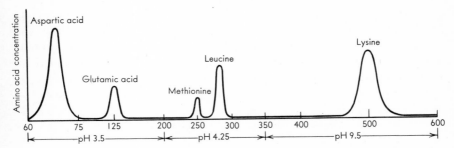

Fig. 47. Separation of amino acids by ion exchange chromatography. The numbers along the base represent the volume of buffer collected from the column.

AMINO ACID SEQUENCE

When, by the analysis of a hydrolysate, we know the proportions of the amino acids in a protein, we can write a general empirical formula of the protein. But to describe the protein's structure more specifically, we must also know the actual sequence of the amino acids in the peptide chain, and whether it is a single or a multiple cross-linked polypeptide strand. The nature of the polypeptide strand is established by both physical and chemical methods. If the protein has only one polypeptide strand, it can have only one free α amino group (NH₂-terminal) and one free carboxyl (C-terminal) group. The free α amino group will react with reagents such as dinitrofluorbenzenes to give a dinitrophenyl (DNP) derivative. After such treatment, the protein can be hydrolyzed and the yellow-colored DNP-amino acid can be isolated by chromatography. If the protein is a pure one and we know its molecular weight accurately, we are in a position to say from the DNP analysis how many NH₂-terminal groups are present. A number of methods are available for determining the C-terminal amino acid residue. The enzyme, carboxypeptidase, is most helpful in this process, because it hydrolyzes peptide bonds that are adjacent to free α carboxyl groups. After treatment with the enzyme, the free amino acid can be analyzed quantitatively by chromatography.

To see if the protein contains more than one polypeptide chain, we split the cross linkages that hold the strands together and then observe whether there is a change in the protein's molecular weight. By unfolding the protein structure, we make the N- and C-terminal groups readily available for chemical reaction.

In Fig. 48, these procedures are briefly summarized. For purposes of illustration, we have used a protein that contains two polypeptide strands linked together by disulfide bonds. If we treat the protein with performic acid, the acid will dissolve the bonds by oxidizing the sulfur, and, consequently, the molecular, or particle, weight will decrease by almost half. If the protein had been single-stranded, the molecule weight would have stayed essentially the same.

By following the above procedures, and applying other chemicals and proteinases such as trypsin and chymotrypsin, Sanger was able for the first time to determine the amino acid sequence in the protein hormone, insulin. Insulin proved to contain two peptide chains, joined by disulfide bonds. In Fig. 49 the amino acid composition and sequence of beef insulin are shown. One chain (A) is composed of 21 amino acids,

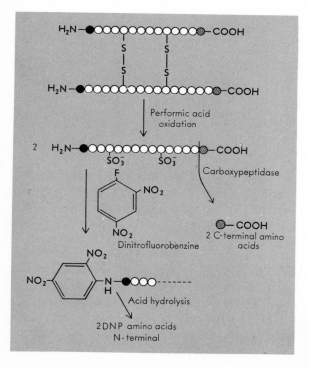

Fig. 48. Determination of N- and C-terminal amino acid residues.

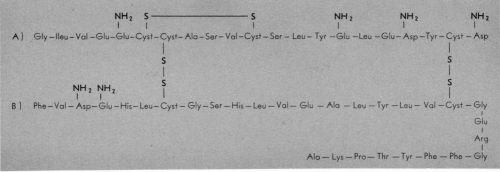

Fig. 49. Amino acid composition and sequence of beef insulin.

with glycine as the N-terminal amino acid and asparagine as the C-terminal. The second chain (*B*) is composed of 30 amino acids, with phenylalanine as the N-terminal and alanine as the C-terminal. The disulfide bridge in the A chain produces a ring structure that involves three cysteine molecules and alanine, serine, and valine. Insulin preparations from other sources differ from beef insulin in the amino acid arrangement of the cyclic structure of chain A. For example, horse insulin contains threonine, glycine, and isoleucine, while swine insulin contains threonine, serine, and leucine.

Secondary and Tertiary Structures of Proteins

Although the peptide bonds, the sequence of amino acids, and the disulfide cross linkages constitute the primary structure of all proteins, these organic chemical characteristics do not explain the behavior of most proteins in solution. Other characteristics, physicochemical in nature, must be attributed to the fact that the polypeptide chain is held in a coiled shape by forces other than those in the primary structure. Chief among these forces is the *hydrogen bond*. This bond results from the tendency of a hydrogen atom to share electrons with two other atoms, usually oxygen atoms. Take formic acid, for example. The early researches of Linus Pauling demonstrated that formic acid exists as a dimer because of hydrogen bonding (Fig. 50). The dotted bonds in the figure indicate that the hydrogen atoms share the electrons of the two oxygen atoms, each of which is in a different molecule, thus pulling the two molecules of formic acid closer together into a stable dimer.

Individually, hydrogen bonds are quite weak, but in a large molecule such as a protein, a number of them will reinforce one another and help stabilize the folded or helical structure. In the polypeptide chain, hydrogen bonds are mainly in the nitrogen and the double-bonded oxygen of the different peptide bonds (Fig. 50). In a typical helical protein, an α-helix, a complete turn contains 3.7 amino acid residues. Since each amino acid occupies the equivalent of 1.47 Angstroms in length along the central axis, every third amide group is linked along the chain by a hydrogen bond.

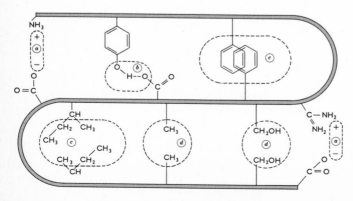

Formic acid dimer   Hydrogen bond in peptide

The helix is further strengthened by other types of non-covalent bonds, in addition to the hydrogen bonds. These interactions constitute the *tertiary structure* of the protein (Fig. 51). Considerable evidence now available supports the view that the secondary and tertiary structures of proteins are essential in biological function.

We have seen how vital water is in biological systems, and it is no less important in protein structure, since it is capable of participating in hydrogen bonds. Water apparently surrounds most protein molecules, through the formation of extensive hydrogen bonds with the protein. To function, proteins apparently must have an aqueous environment.

One final point may be made with regard to the tertiary structure of proteins. The availability of possible cross linkages between proteins will increase as the degree of coiling (and internal bonding) decreases. Therefore, fibrous proteins (long and relatively uncoiled) will tend to interact with one another through various weak linkages, and, as a result, solutions of proteins will tend to be more viscous than solutions of other substances. The cytoplasm of the cell is a viscous colloidal system for this reason. Furthermore, the degree of interaction will be dependent upon the environmental conditions, since changes in pH, temperature, salt concentration, etc., will affect the nature or the availability of these cross-links. The observed changes in shape, viscosity, streaming, and so forth, characteristic of cells like the amoeba or cells in division, can be understood in terms of this changing colloidal system, called a *thixotropic gel*. The study of the submicroscopic structure of cells has become one of the areas of active investigation during recent years. Although we cannot deal with this subject here, it is important to note that many of the properties of cytoplasm can be explained in terms of the secondary and tertiary structure of protein molecules.

Fig. 51. Some types of noncovalent bonds which stabilize protein structure: (a) Electrostatic interaction; (b) hydrogen bonding between tyrosine residues and carboxylate groups on side chains; (c) interaction of nonpolar side chains caused by the mutual repulsion of solvent; (d) van der Waals interactions. (Christian B. Anfinsen, *The Molecular Basis of Evolution*, John Wiley & Sons, Inc., 1959.)

An enzyme is a protein that changes the rate of a chemical reaction but does not affect the nature of the final products; in other words, it acts as a catalyst. It would be incorrect, however, to infer from this definition that the enzyme does not participate in the reaction, for as we saw when discussing the mechanism of action of the glyceraldehyde-3-phosphate dehydrogenase, the SH group in the protein definitely combines with the substrate. As the reaction proceeds, however, the enzyme is liberated.

Catalysts may be characterized by the following properties. (1) They are effective in very small amounts. When analyzing the catalytic power of an enzyme, we should know the amount of starting material or substrate that is converted to product in a unit time by a given quantity of the enzyme. This quantity is called the *turnover number* and is defined as the number of moles of substrate converted into product per minute by one mole of enzyme. The turnover numbers of enzymes vary greatly, ranging from one hundred to over three million. (2) Catalysts are usually unchanged in the reaction. This property of ideal catalysis can only be approximated by enzymes, however, since they are not completely stable under the conditions of most experiments. Proteins are extremely unstable substances and are easily inactivated or denatured by high temperatures or very alkaline or acidic conditions. An extreme example of denaturation occurs when an egg white is boiled to harden it. (3) Catalysts ordinarily have no effect on the equilibrium of a reversible chemical reaction; they merely speed up the reaction until it reaches equilibrium. The function of the true catalyst, therefore, is to hasten the process in either direction. (4) Catalysts exhibit specificity in their ability to accelerate chemical reactions; that is, a given catalyst affects only certain types of chemical reactions. We shall speak of this specificity at greater length later.

The most dramatic advance in biochemistry during the last thirty years has been the isolation of a variety of enzymes from different cells, thus enabling us to study their mechanism of action in solution away from the complications of cellular activity. The complicated reactions involved in carbohydrate, fatty acid, and amino acid metabolism have been analyzed step by step. The process of isolating enzymes from cellular extracts and purifying them has been a slow but a rewarding one.

On the basis of an impressive body of information, we now believe that enzymes are definitely proteins. To understand how enzymes exert their catalytic action, therefore, we must know the details of protein structure. Since large gaps still remain in our knowledge of this structure, many aspects of the mode of action of enzymes are still obscure, but this does not bring us to a halt, for the properties of enzymes as catalysts may still be studied without immediate regard to the mechanism of action. From such investigations, valuable data about how enzymes act in a biological system have been obtained.

Effect of Temperature on Reaction Rate

As we have said, the function of an enzyme is to increase the rate of a chemical reaction. In water, the molecules of the cell are in ceaseless thermal motion, and occasionally react when they collide with one another. In a solution without enzymes, the chance that a molecular collision will result in reaction is very small; but if the appropriate enzyme is present, the probability will be greatly increased. The major question is: How do enzymes perform this unusual catalytic activity? The answer seems to lie in the ability of enzymes to make a substrate molecule more labile or more reactive to other molecules in the environment.

Chemical compounds that can be isolated are more or less stable, for unstable molecules, by definition, react with other molecules in the environment until they form more stable products. Stable molecules can react rapidly, however, if they are activated by the addition of energy. Arrhenius was the first to point out that all the molecules in a given population do not have the same kinetic energy. Some molecules, through collisions, acquire more energy than others, and these energy-rich molecules are more likely to react than energy-poor ones. In other words, there is an energy barrier to reaction, and the higher the energy barrier for a molecule, the greater is its stability. The energy required to hurdle molecules over this barrier is called the *energy of activation.*

Figure 52 charts a hypothetical reaction in which A is converted into B. This diagram holds for all chemical reactions, although the height of the energy barrier varies from one reaction to another. Note that for A to be converted into B, it must first acquire the necessary energy to form the activated molecule $A^{\ddagger}$. The inherent rate of a chemical reaction depends on the number of $A^{\ddagger}$ molecules existing at any one moment and the frequency of their decomposition (C) into products, B; or:

$$\text{rate} = (A^{\ddagger})C$$

It turns out that C is a constant and is essentially the same for all chemical reactions. In order to describe the absolute rate of a chemical reaction, we must find a way to determine the concentration of $A^{\ddagger}$. It can be shown that the number of activated molecules is proportional to the energy of activation (E), and consequently the rate of the reaction can be written as: rate $= Ce^{-E/RT}$, where $e^{-E/RT}$ is proportional to the number of molecules in the excited state. We determine E by studying the rate of reaction at different temperatures. When we plot the logarithm of the rate against the reciprocal of the absolute temperature (C' + 273), we obtain a straight line whose slope equals E/R.

It is evident that as the temperature is raised, the rate of the reaction increases logarithmically. As a general rule, for every 10°C rise in temperature, the rate of chemical reactions increases 2–3 times (Fig. 53). The rise in temperature increases the number of activated molecules by increasing their movement and the number of collisions. The rotation and vibration of molecules also increase with rising temperatures, and all these processes produce a greater proportion of molecules that have the necessary activation energy for reaction.

What has all this to do with enzyme-catalyzed reactions? One of the significant features of enzymes is their ability to lower the energy of

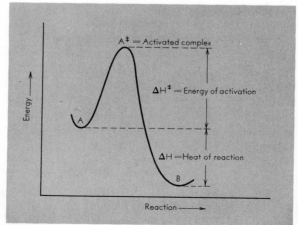

Fig. 52. Energy of activation for a chemical reaction.

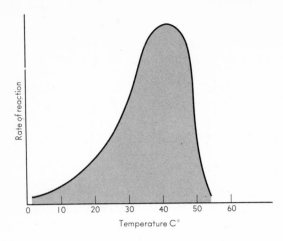

Fig. 53. The effect of temperature on an enzyme-catalyzed reaction.

activation, enabling reactions to occur at ordinary physiological temperatures. Exactly how they do this we do not know. Reactions that might ordinarily occur only at boiling temperatures, because of the energy barrier, can now take place with relative ease in the cell.

No one has yet succeeded in isolating an activated complex. Activated complexes have only been deduced from kinetic studies of enzyme-catalyzed reactions carried out at different temperatures. Temperature affects ordinary chemical reactions in the same way it affects enzyme-catalyzed reactions. At high temperatures, however, the heat tends to destroy (denature) enzymes so that no further reaction can take place. Biological phenomena thus have optimal temperatures at which they function. Above a critical temperature enzymes do not perform. As shown in Fig. 53 for a hypothetical situation, the reaction rate increases rather rapidly and logarithmically from 0° up to about 25°C, but above this point, the increase slows, and at approximately 35°C the rate starts to decrease. If the damaging temperature is maintained for too many minutes, the enzyme will be completely inactivated.

The thermal behavior of enzymes imposes serious limitations on organisms; most cells lose their capacity to carry on metabolism at temperatures above 40°C. Only the few organisms that have heat-resistant enzymes are able to survive in exceptionally hot places. In most cells, the rate of metabolism, and hence the intensity of the life processes, changes as the temperature varies from day to day and from season to season. The winter metabolism of most organisms subsides to a point where dormancy is inevitable. Warm-blooded organisms such as man and a few other vertebrates are the only creatures that have evolved a method of controlling their body temperature.

Enzyme-substrate Complex

As is probably evident from our previous discussions, enzymes do more than lower the energy of activation. They seem, in some cases, to set the pathway or direction in which a particular substrate will be metabolized. For metabolic processes to occur, the substrate molecule must come into intimate contact with the specific enzyme. By combining with the enzyme, a deformation in some of the bonds of the substrate molecule probably occurs, thus producing a condition that favors the reaction.

The idea that the substrate must combine with the enzyme before catalysis can proceed is an old one, and there is much experimental evidence to support it. If the substrate molecules must combine with the enzyme before catalysis takes place, collision, then, must play an important part in the process. Since the enzyme molecule is very large and the substrate molecule is usually quite small, for the substrate to combine with the enzyme there must be a relatively large number of substrate molecules in order to increase the probability of correct collision.

When we study the rate of an enzyme-catalyzed reaction involving different amounts of substrate molecules, we observe that the reaction rate does indeed vary with the concentration of substrate, at least over a limited range of substrate concentration (Fig. 54). The rate curve at first rises, then slopes off, and finally reaches a constant maximum value. Where the rate of the reaction no longer changes with an increase in the substrate concentration, we assume that the enzyme surface is completely covered or saturated with substrate. Two different enzyme concentrations are shown in Fig. 54. Note that the maximum rate (Vm) depends on the enzyme concentration.

The combination of a substrate with an enzyme is a very specific process, i.e., each catalyst has a unique surface containing a precise spot where the substrate molecule can join it. Because of the unusual folded surface of proteins, only very specifically shaped molecules can gain access to the particular chemical groups (such as SH). For this reason, the enzyme-substrate complex is often compared to a lock and key, as shown in Fig. 55, where the substrate is analogous to the key. We can

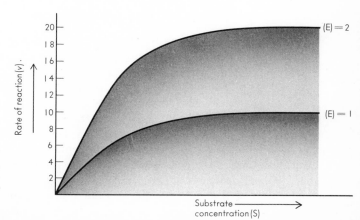

Fig. 54. Effect of substrate concentration on the rate of enzyme catalyzed reactions.

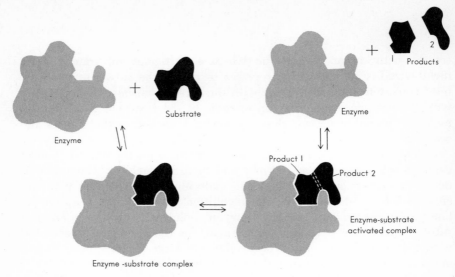

Fig. 55. The formation of an enzyme-substrate complex, followed by catalysis.

write the reaction which describes Figs. 54 and 55 in the following way:

$$E + S \underset{k_2}{\overset{k_1}{\rightleftharpoons}} ES \overset{k_3}{\longrightarrow} E + products$$

Here k_1 refers to the rate of combination of E and S to form ES, and k_2 is the rate for the dissociation of ES to E and S. After the substrate combines with the enzyme to form ES, we assume that enough activation energy is acquired so that ES is converted into the product of the reaction, which dissociates from the enzyme surface to allow the enzyme to combine with another substrate molecule. The rate of conversion of ES to the products is indicated by the constant k_3. At a low substrate concentration some of the enzyme is free, and the maximum rate is not observed. With excess substrate, all the enzyme is converted into ES, and the reaction takes place at its maximum rate—i.e., maximum velocity $(Vm) = k_3(ES) = k_3(E)$.

Enzyme Inhibitors

As we indicated previously, the fact that an enzyme-substrate combination is a specific one has led us to believe that a particular enzyme and its substrate must possess shapes that complement one another if they are to fit together. In other words, the general configuration of the protein molecule must be such that the substrate molecule can fit into a specific site on the enzyme. The specificity and geometry of enzyme reactions are beautifully illustrated by the phenomenon of *competitive inhibition.* To demonstrate this concept, let us consider the enzyme, succinic dehydrogenase, which catalyzes the oxidation of succinic acid. If malonic acid, whose structure closely resembles that of succinic acid, is added to the solution, the enzyme's effectiveness is considerably reduced. Experiments show that malonic acid, while not manifesting any change, apparently attaches itself to the enzyme at a position on the molecule that would normally be filled by succinic acid, and therefore competes with succinic acid for the active region of the enzyme molecule.

In Fig. 56, the enzyme and substrate are able to achieve an exact "chemical fit," as in the lock and key analogy we have already mentioned. The inhibitor, malonic acid, has a structure similar to that of the substrate and is able to combine at least with part of the active site on the enzyme, thus preventing the substrate from entering. In this way, the enzyme is removed from its catalytic role. Since substrate and inhibitor compete for the same site, the degree of inhibition depends on the ratio of substrate to inhibitor. If there is an excess of substrate, the inhibitor may not be able to compete, and no inhibition will be observed. We use this fact to determine whether a substance is acting as a competitive inhibitor; that is, we increase the substrate concentration until the inhibition is removed. Some allowance must be made, however, for the relative affinities of the substrate and the inhibitor for the enzyme. For instance, to obtain a 50 per cent inhibition of an enzyme reaction, we may have to put into the reaction mixture more inhibitor molecules than substrate molecules, for the affinity of the inhibitor for the enzyme may be less than the substrate's affinity for the enzyme.

Competitive inhibitors abound, and some of them are effective antibacterial agents. One of these is sulfanilamide (Fig. 57). After the chance discovery that this substance inhibits the growth of certain bacteria, careful analysis revealed that the B vitamin, p-aminobenzoic acid (PABA, Fig. 57), removed the inhibition. Subsequent studies indicated that this was an example of a competitive inhibition.

The success of sulfanilamide and other bacteria killers opened the exciting possibility that all disease-producing organisms, and cancer as

Fig. 56. Competitive inhibition.

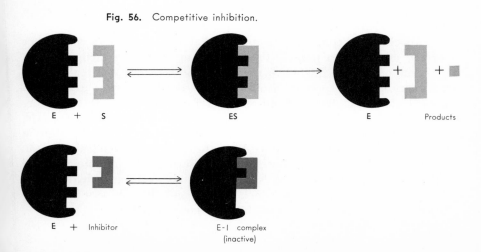

E   +   S   ES   E   Products

E   +   Inhibitor   E-I   complex
                                              (inactive)

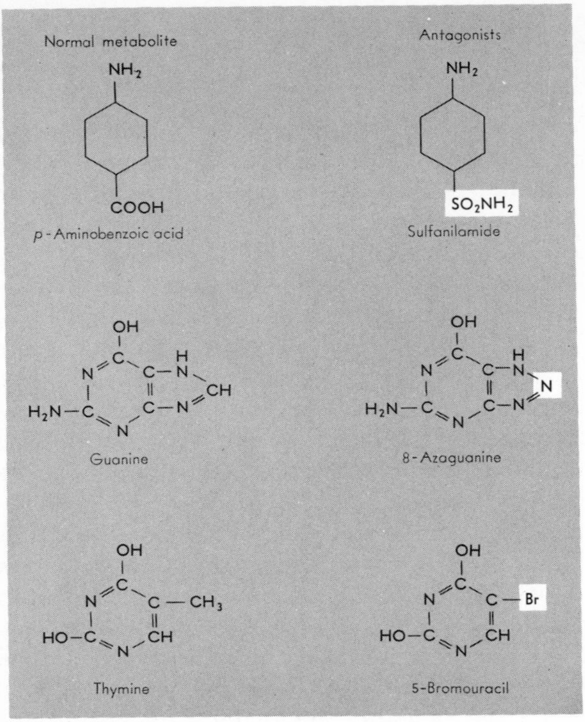

Fig. 57. Metabolic antagonists.

well, might be susceptible to the antimetabolite approach, but, unfortunately, the host and disease-producing organisms have very similar enzyme machinery, which means that the altered vitamin or substrate is also quite toxic to the host cells. Since there are some differences, however (in rates, affinities, etc.), between the host and its enemy's enzyme-catalyzed reactions, by careful screening we may be able to find the appropriate chemical that will kill the disease before harming the host cells. PABA, for example, is an essential part of folic acid, a key coenzyme for the transfer of methyl ($-CH_3$) groups in both animals and plants. Why certain bacteria are more sensitive to the sulfa drugs than are the host cells is not fully understood at present, but this is a case in which such host-enemy differences exist.

Biochemists commonly employ metabolic antagonists and other enzyme inhibitors to unravel the mechanism of enzyme action. When we speak of a "close fit" between enzyme and substrate, we really mean that there must be specific chemical groups on the enzyme arranged in a particular way. Research efforts, therefore, are directed to finding chemicals that will specifically combine with certain groups on the enzyme. If these groups are important for catalytic activity, inhibition will be observed. In Table 4 we list a few of these chemicals and an example of an enzyme which each inhibits. Fortunately, in some cases the inhibitor-enzyme combination is stable, and so we can, by hydrolyzing the protein, determine with which amino acid the inhibitor has combined. Obviously, this type of inhibition is not competitive.

Note that many of the reagents listed in the table are normally classified as poisons. Compounds such as cyanide and carbon monoxide, by combining with the cytochromes, completely block the electron transport to oxygen. In recent studies, diisopropylphosphofluoridate (Fig. 58) has been successfully used as an enzyme inhibitor. For example, when the proteinase, chymotrypsin, combines with DPF, one molecule of inhibitor combines with one molecule of enzyme, causing a complete inhibition of catalytic activity. By degrading the protein-inhibitor com-

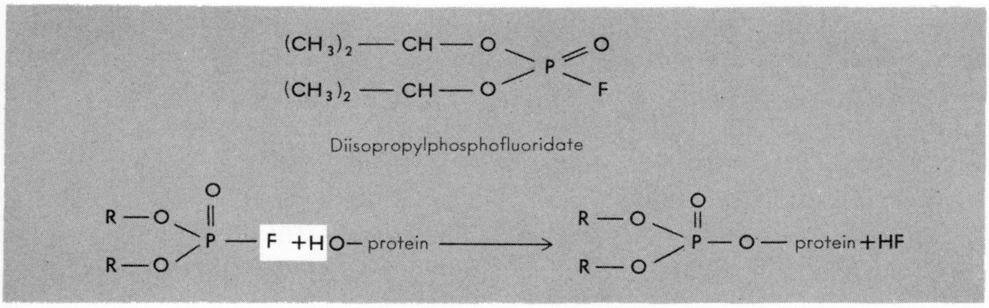

Fig. 58. Inhibition of an enzyme by a specific group reagent.

plex, we can isolate a very small peptide (6 amino acids) to which the DPF is attached. Such studies seem to indicate that serine and histidine are two amino acids associated with the catalytic site. Interestingly, recent studies have shown that five enzymes with quite different catalytic activity have these same six amino acids associated with the catalytic site.

Table 4
Enzyme Inhibitors

Inhibitor	Group on enzyme affected	Example
p-chloromercuribenzoic acid	—SH	Urease
Iodoacetic acid	—SH and others	Triose phosphate dehydrogenase
Mercury and silver	SH and anions	Glutamic dehydrogenase
Cyanide and azide	Metalloporphyrins and metals	Catalase
Carbon monoxide	Metalloporphyrin	Cytochrome oxidase
Citrate, oxalate	Removes metals	Hexokinase
Diisopropylfluorophosphate	Combines with OH group of serine	Chymotrypsin

The recent, exciting studies of the mechanism of enzyme action are concentrating on the specific reactive groups on the protein molecule and their relationship to the neighboring amino acids, making it imperative that we know the amino acid sequence in the protein as well as the chemical nature of the active site. In some cases, it has been possible to remove some of the amino acids from the polypeptide without affecting its catalytic activity, providing further evidence that there is an active site and that the entire protein is not essential for catalytic activity.

Catalytic activity is sometimes lost when a small peptide is removed from the protein, but it can be restored to the protein residue by adding equal amounts of the small peptide to the reaction mixture. Since no peptide bonds are formed under these conditions, hydrogen bonding and other non-covalent linkages must be essential in restoring the catalytic site, a fact that reinforces the theory that the secondary and tertiary structures of protein play a key role in the functioning of enzymes. It will be up to future studies to determine whether any loss in specificity results when part of the polypeptide chain is removed or altered.

Types of Enzymes

From our earlier discussion, we can see that there are many different types of enzymes and that the mechanism of action of each type must be investigated. Some enzymes require the presence of compounds of low molecular weight before they are active, and it follows that these compounds must be partly responsible for producing the active site of the enzyme. On the other hand, chymotrypsin is a hydrolase that requires no cofactors and catalyzes the following type of reaction:

$$\underset{\text{(enzyme)}}{\text{EH}} + \underset{\text{(substrate)}}{\text{RX}} \longrightarrow \underset{\text{(intermediate)}}{\text{E} - \text{R}} + \text{XH}$$

$$\downarrow \text{H}_2\text{O}$$

$$\text{EH} + \text{ROH}$$

$$\text{over-all: } \text{RX} + \text{H}_2\text{O} \longrightarrow \text{ROH} + \text{XH}$$

In many enzyme-catalyzed reactions of this type, water is the special chemical that breaks the bond. The protein's configuration gives it the specificity for particular substrates. Most protolytic enzymes, lipases, and phosphatases fall into this category. A modification of this type of reaction is of particular interest; in it the enzyme catalyzes the following kind of transfer reaction. The enzyme, E, combines with the A portion of AB to form $\text{E} \sim \text{A}$, thus conserving the bond energy of the AB link which can then react with C to form AC. The important point here is that water does not react with the $\text{E} \sim \text{A}$ complex, so the energy of the bond is conserved and used to synthesize a new compound, AC. This is the case with such enzymes as hexokinase, myokinase, CoA transfer reactions, and carboxylase.

One way of testing this proposed mechanism is to place some

isotopically (radioactive) labeled B* in a reaction mixture along with AB and allow the reaction to proceed in the presence of the enzyme. After a suitable time, but before all the AB is broken down, one should find the radioactive label in AB*, indicating that an enzyme~A complex is formed that is free of B:

$$E + A{-}C$$
$$\uparrow + C$$
$$E + A{-}B \rightleftharpoons E \sim A + B$$
$$\downarrow\uparrow + B^*$$
$$E + A{-}B^*$$

In other words, in the absence of C the enzyme forms E~A and B; however, since the reaction is reversible B can immediately react with the E~A to reform AB and a free enzyme. If we place some labeled B* in the reaction mixture, it can compete with unlabeled B in the reverse reaction, thus forming the labeled substrate A—B*.

If labeled A* is used in place of labeled B, however, no label will be found in AB. Such isotope-exchange reactions have been employed to study the mechanism of action of a number of enzyme-catalyzed reactions. In evaluating the data obtained in this manner, we must consider the purity of the enzyme preparation and be certain that the over-all reaction is the one that is being described. Also important in this type of reaction is the fact that the molecule attached to the enzyme surface is restricted in reactivity. For example, AB might react with a large number of molecules if it were not hindered by the enzyme. In addition, the E~A complex can only react with molecules that have a configuration similar to B, as is beautifully illustrated in the following reaction, in which several active sites must exist on the enzyme (E).

Before free acetic acid can be metabolized in the citric acid cycle, the acid must be converted into acetyl coenzyme A according to these reactions:

$$E + \text{acetic acid} + ATP \rightleftharpoons E\text{-acetyl-AMP} + PP$$
$$E\text{-acetyl-AMP} + CoA \rightleftharpoons E + AMP + \text{acetyl-CoA}$$

The incorporation of radioactive pyrophosphate (PP^{32}) into ATP and the synthesis of ATP from acetyl-AMP and PP demonstrate that the first reaction is reversible. Normally, acetyl-AMP (as well as other acyl-AMP compounds) is very unstable and will react with a number of compounds, but when it is bound to the enzyme, CoA is apparently the only compound of biological importance that will react with it. When the reactive intermediate is bound to the enzyme, therefore, it is restricted to a given metabolic pathway.

Chemical change does not necessarily take place simply because a substrate molecule has fitted onto the protein molecule, for the presence of other substances that are not proteins may be required before a reaction occurs. The cytochrome proteins, for example, have such accessory substances (the iron porphyrins) more or less permanently attached to them, and these tightly bound substances, called *prosthetic groups,* are essential for the electron-transport process. In order to function, many enzymes must have an accessory substance which readily dissociates from it; these accessory substances are called coenzymes, and we have already discussed their role in this kind of reaction.

In addition to the organic cofactors (vitamin-containing compounds), a number of inorganic cofactors are required for enzyme catalyses. For example, iron porphyrins, as we know, are necessary for electron transport, but iron also seems to be essential for electron transport in most organisms even when it is not bound in an organic form. Copper, too, apparently functions in electron transport; magnesium is essential for the transfer of phosphate groups from ATP, and so on. Examples of these required metals are presented in Table 5. The mechanisms of action of these various cofactors constitute one of the most active fields in enzyme research today, but much is yet to be learned. It is clear, however, that in some cases the enzyme protein alone is not enough to speed the chemical reactions.

Table 5
Some Enzymes Whose Activity Depends on Metals [1]

Metal	Enzymes
Magnesium	Phosphatases, ATP reactions, chlorophyll
Copper	Tyrosinase, respiratory proteins in invertebrate animals
Zinc	Various dehydrogenases, peptidases, carbonic anhydrase
Iron	Cytochromes, hemoglobin, electron-transport systems in mitochondria
Manganese	Peptidases, some enzymes of the citric acid cycle
Cobalt	Peptidases, in vitamin B_{12}
Molybdenum	Nitrate reduction, xanthine oxidase
Potassium	Phosphopyruvate transphosphorylase, fructokinase
Calcium	Actomyosin

[1] This is a very limited list. Additional metal activated enzyme systems may be found in advanced textbooks or monographs.

Control
of Cellular
Metabolism

In order to function, a living cell must rely on a complex series of reactions. The glycolytic and oxidative cycles, and fatty acid and amino acid metabolism, are examples of interdependent activities which affect all cells. In addition, cells that have specific functions must channel their major energy expenditures through special pathways that enable them to perform their particular tasks. To regulate and to coordinate the multienzyme systems of the cell so they will achieve a specific final result require precise control mechanisms. One of the greatest challenges facing the physiologist today is that of piecing together our knowledge of the active framework of the living cell, the framework that supports the ordered processes of life.

From our earlier considerations of cellular metabolism, it is evident that both the rate of reactions and the direction they will take is dependent on the relative concentrations of substrate and enzyme molecules. We must consider the cell as a vast equilibrium system, in which control can be achieved only through alterations in the concentrations of the reactants. In a very crude extract of yeast cells, which will ferment carbohydrate in a reasonably ordered manner, the rate of carbon dioxide production (and of alcohol synthesis) can be controlled by the alteration of the concentrations of such reactants as sugar, phosphate, ATP, DPN, etc. Obviously, within the intact cell, the same factors will exert controlling influences. For example, when a muscle cell contracts suddenly, ATP is broken down, forming inorganic phosphate and ADP, both of which are essential in the stimulation of carbohydrate breakdown. Inorganic phosphate stimulates the phosphorolysis of glycogen to form glucose-1-phosphate, while ADP tends to remove phosphorylated compounds in the triose phosphate dehydrogenase reaction. In this way, the muscle cell can "know" when it needs energy for contraction and relaxation and can supply that need.

In addition to such substrate effects, evidence accumulated over the past few years indicates that the reactants (substrates, enzymes, etc.) do not mix as readily in the cell as they do in the test tube. The reason for this is that, within the cell, enzymes may be bound into the various internal structures so that enzymes and substrates are spatially separated. We have already seen that the mitochondria are cellular units which contain the enzymes concerned with the oxidative metabolism of carbohydrates, as well as with the metabolism of fatty acids and amino acids. In several cases that have been closely studied, the enzymes that are functionally related are apparently tightly bound together within the structure of the mitochondrion. Thus the enzymes of the Krebs citric acid cycle and those concerned with oxidative phosphorylation seemingly need to be in close proximity for proper functioning. On the other hand, the mitochondrial membrane is not permeable to many substrates, which insures a separation between the oxidative and non-oxidative metabolism of carbohydrates.

Many substances of small molecular weight, such as ATP, inorganic phosphate, coenzyme A, and acetylcholine, are also tightly bound and highly localized in cells. Since enzyme systems and their cofactors which are fixed to insoluble units of the cell may be of primary significance in the regulation of cellular activities, we cannot limit our discussion of the control of cellular metabolism to the enzyme activity we view in the test tube. We must also examine the structural organization of enzyme systems—in other words, the cellular architecture. However, the localization and separation can only affect the relative concentrations of reactants, so we are safe in saying that variations in metabolic activity caused by changes in this structural organization result from differences in either the enzyme concentration or enzyme activity.

It must be obvious from this discussion that those compounds that are in the lowest concentration and in the greatest demand will be the ones that control the metabolism with the greatest sensitivity. The low concentrations of cofactors such as ATP and DPN provide a method for controlling both the rate and the direction of the entire metabolism, although the depletion of even these factors takes more time than is required for a cell to respond to certain environmental stimuli. We can thus term such mechanisms "slow responding" controls.

In addition to altering the availability of substrates and cofactors, reactions can be controlled by changing the total amount of active enzyme (as contrasted to the distribution of enzyme). When we compare the enzyme activity of cells under different physiological states, we must determine whether observed increases or decreases are causd by alterations in the activity of pre-existing enzyme molecules or by the synthesis of new molecules. The time interval involved is crucial in such studies. A change that occurs in a short time is of great diagnostic value, since it most likely indicates enzyme activation or inhibition. To explain the very fast off-and-on effects produced in cells by stimulation, we must go beyond the simple fact that the enzyme and substrate are isolated from one another. Minor structural alterations, which convert an enzyme inhibitor into an inactive form or vice versa, may be responsible for the faster mechanism.

At the present time, we know of many instances in which enzymatic activity can be demonstrated in a cell-free extract only after an inhibitor has been removed. Some of these inhibitors are actually proteins themselves and can be inactivated by heat. The products of an enzyme-catalyzed reaction can also act as inhibitors if they are tightly bound to the enzyme; in this case, the enzyme is enzymatically inactive until the product is removed. The products of a reaction, therefore, may be important in the control or regulation of metabolic processes. Although we know that rapid cellular responses are produced when an inhibitor is removed from an enzyme surface, much work remains to be done on this type of control mechanism before we completely understand it. This area of investigation assumes extreme importance when one considers that the action of hormones probably falls into this category.

Ultimately, the control of cellular metabolism must be accomplished through the regulation of enzyme synthesis. The constitutive enzymes of a cell are the main determinants of the cell's form and function. Although cells have a remarkably wide range of responses to the environment, cells of the same type show similar properties and similar responses. For this reason, it is clear that the hereditary mechanisms of cells are fundamental to the control of cellular activity. Modern genetics holds that the primary direction of the elaborately interrelated processes of metabolism, development, and function come from genetic material carried in the chromosomes of the cell nucleus. The nature of this material, its method of production, the manner in which it functions, the mechanisms of its transmission from one generation to another, and its role in the process of organic evolution are basic factors in the science of genetics. Although another book [1] in this series considers the detailed aspects of genetics, we must take up some of these problems here, since the control of cell division, the duplication of the genetic material, and the synthesis of the metabolic units of the cell are important activities in biology that are of primary interest also to physiologists and biochemists.

---

[1] D. M. Bonner, *Heredity* (Englewood Cliffs, N.J.: Prentice-Hall, 1961).

Geneticists and cytologists have established that the hereditary material in all organisms is basically the same, and careful studies have revealed that this material is transported in the chromosomes. In a particular chromosome, each of the many genes appears to occupy a fixed and special position known as a locus. The order and spacing of the genes in a single chromosome are determined by recombination and linkage tests (which are described in the book on the cell in this series).[2] That these positions are correct and are located in the chromosomes can be verified by correlating visible chromosome changes with certain biochemical deficiencies. For example, the absence of a specific small segment of chromosome can sometimes be related to the absence of a specific gene which controls a known biochemical step.

One method of finding out about genes is to study their mutational properties. From such studies, Beadle and Tatum concluded that a gene mutation alters cellular processes by affecting protein synthesis. When it was later found that gene mutation changes the nutritional requirement of an organism, it became obvious that the genes control the biosynthesis of the basic chemical units of the cell by controlling the synthesis of specific enzymes that are the essential catalysts for biosynthesis.

For example, in Fig. 59 is outlined a series of hypothetical reactions which lead to the synthesis of the amino acid, histidine. Each biochemical step in this synthetic pathway is catalyzed by a specific enzyme (E_A, E_B, etc.). The synthesis of these enzymes, in turn, is under the control of specific genes (G_A, G_B, etc.) in the chromosome. Mutational studies have shown that if we alter gene A we also affect the synthesis of enzyme A and thus prevent the immediate synthesis of B, which is an essential precursor of histidine. If an organism experienced such a mutation, it would require histidine in its diet in order to survive and grow.

Since the hypothesis that genes control enzyme synthesis is reasonably supported by experimental facts, we are confident today that genes exert their control largely through the enzymes they produce. In an early chapter, we considered the wide variation that exists in the nutritional requirements of organisms. It is now clear that this tremendous variation is caused by changes in the genetic constitution of the

---

[2] C. P. Swanson, *The Cell* (Englewood Cliffs, N.J.: Prentice-Hall, 1960).

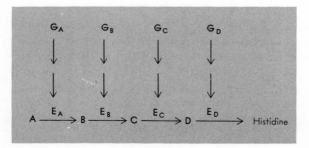

Fig. 59. The genetic control of histidine biosynthesis.

organisms. The ability of an organism to respond to a particular environment and to gain a specific metabolic ability, therefore, depends on its genetic constitution. For example, certain strains of *E. coli* cells, when grown on glucose and ammonia as the primary carbon and nitrogen sources, are capable of making a number of enzymes that normally do not occur in the cells. One such enzyme is β-galactosidase, which catalyzes the hydrolysis of β-galactosides. A typical substrate is the disaccharide, lactose, which occurs in milk. On hydrolysis, lactose gives glucose and galactose, and when it is added to yeast cells grown on glucose, there is a long lag before the cells can metabolize it.

A number of experiments support the conclusion that during this lag period the enzyme β-galactosidase is synthesized by the cells, indicating that specific substrates in the environment can often *induce* the formation of new and specific enzymes that are essential for the metabolism of these substrates. Such enzymes are called *inducible enzymes,* in contrast to the *constitutive enzymes* that are normally present in the cell. Subsequent studies have shown that the cells must have the appropriate genetic constitution before they can respond to a substrate and make a specific enzyme. The ability of an external substrate, in other words, to induce the formation of a particular enzymatic pattern depends on the genetic constitution of the organism.

This knowledge is a tremendous step toward our general goal of decoding the inherited message in the nucleus. We are now reasonably certain that the message is chemical in nature and that the genes, through their chemical composition and internal arrangement, control the protein-forming center. Perhaps it is the "chemical surface" of the gene that is the key factor in this control; as with enzymes and substrates, the gene surface may in some way be complementary to the enzyme surface. The gene's ability to produce an exact copy of itself in each cell generation is its most outstanding property, and this ability could well be explained by its capacity to make an enzyme, thus imparting its character to the enzyme as a mold fashions the shape of a casting. Genetic control of metabolism, therefore, is exercised primarily through the ability of genes to regulate, in some way, the synthesis of proteins with specific primary and secondary structures. We now investigate the possible ways the chemistry of the genetic material may control protein synthesis.

The Chemistry of the Genetic Material

Chromosomes consist mainly of two kinds of chemical substances, protein and deoxyribose nucleic acid (DNA). A second type of nucleic acid, ribose nucleic acid (RNA), also exists in small concentrations in the nucleus, and much larger quantities are found in the cytoplasm. DNA synthesis and chromosome replication are closely connected with the general reproduction process at the subcellular level, and we are confident that the DNA in the chromosome is the chemical substance primarily concerned with the transfer of hereditary information from one generation to the next. Later we shall discuss a number of experiments that support this theory.

When DNA is isolated from almost any source, it is found to consist of thread-like particles of very high molecular or particle weight (several millions). Its chemical composition varies greatly, but these differences seem to be due largely to the arrangement of a few simple chemical compounds called purine and pyrimidine bases. So far about ten different bases have been found in nucleic acids, although four of these bases (2 purine and 2 pyrimidine) seem to occur more frequently and in higher concentrations than the others. In the intact nucleic acid molecule, these bases are attached to a five-carbon sugar, called 2-deoxyribose, to form a deoxynucleoside. The nucleosides are linked together by a phosphate group at the number 3 position of one nucleoside and the 5 position of a second (Fig. 60).

When DNA is hydrolyzed by acid, the first products observed are deoxyribonucleotides. As indicated in Fig. 60, the deoxyribose and phosphate form the chemical backbone of the DNA molecule, and the

Fig. 60. DNA chemistry.

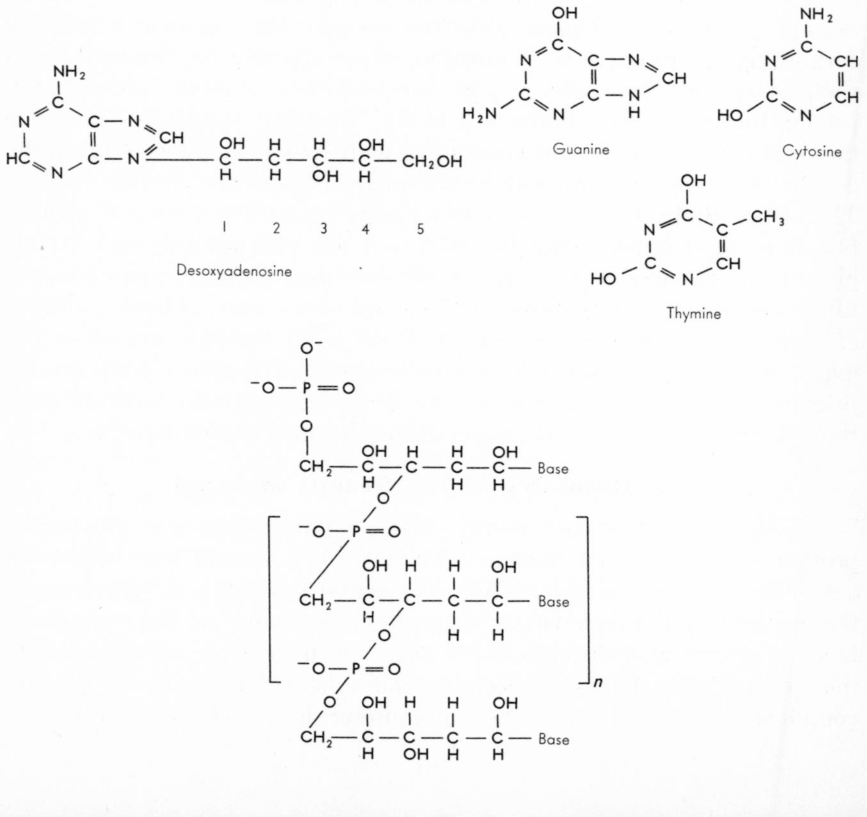

variation in chemical composition must depend on the purine and pyrimidine bases. Present evidence strongly suggests that the sequence of these bases must be the major factor in determining the functional (hereditary) properties of the DNA molecule. Chemical analyses of a number of samples of DNA indicate that the bases do not occur in a random fashion and that the sum of the purines equals the sum of the pyrimidines; the content of adenine also equals that of thymine and the content of guanine equals that of cytosine. In some DNA molecules, the A plus T content is much larger than the G plus C content, while in other preparations the reverse is true; the significance of these differences is not understood. However, if we are to explain the diversity of molecules that are under genetic control, we must have a system in which different "surfaces" may be set up with some facility. Since the possible combinations of nucleotides in the DNA molecule are of the order of 10^{200}, this molecule has great potential for the transmission of information in the form of a molecular code to future generations. The potentialities of DNA differences are also important when we consider that each species of organism has a different base sequence. The order, however, must be characteristic of the species. Since there are well over two million species of organisms, there must be at least that many different DNA molecules whose identity depends only on base sequence.

With this information about the restrictions on base pairing and with additional physical data on DNA molecules, Watson and Crick in 1953 constructed a three-dimensional model of DNA. Its uniform molecular pattern is produced by two polynucleotide chains arranged in a helical structure, and the two chains are held together by hydrogen bonding between the bases. Chemical data and the symmetry of the helix suggested that adenine-thymine and guanine-cytosine were the base pairs and that the bases on one strand determine the sequence on the partner strand. The hydrogen bonding of these pairs is shown in Fig. 61. It is evident from this pairing that if the bases in one strand are arranged in the order GAGGTC, the complementary strand will be CTCCAG (Fig. 62). This conception of the DNA molecule is one of the major achievements of biochemistry. Although the genetic significance of the base sequence is not known, this working model of the DNA molecule has stimulated a number of new and interesting experiments.

From a biological viewpoint, the proposed DNA structure is quite satisfying, since it is one of the few models so far constructed which reasonably explains how a complex and highly specific molecule can be duplicated from an array of building blocks. With this model, we assume that during the replication process, the complementary polynucleotide chains separate and that each one acquires a new partner from the base pool. Experiments using the heavy isotope of N^{15} indicate that when new DNA strands are produced there is no degradation of the old strand, a result that is in keeping with the idea that the old DNA strand acts as

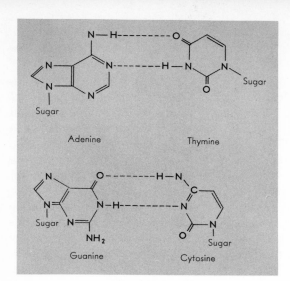

Adenine

Thymine

Guanine

Cytosine

Fig. 61. Pairing of purine and pyrimidine bases by means of hydrogen bonding.

Fig. 62. A schematic diagram of the DNA molecule showing the base pairing and the helical structure.

a template on which a new strand is then made.

Recent experiments by Kornberg and his colleagues on the enzymatic synthesis of DNA-like molecules suggest that the base pairing of the new material is determined by the small amount of DNA present initially. This cell-free synthesis requires the presence of the corresponding triphosphates of the purine and pyrimidine bases (i.e., deoxy ATP, TTP, GTP, and CTP), the appropriate enzyme preparation, and a DNA primer or starter. The enzyme catalyzes the condensation of the deoxyribonucleoside triphosphates, liberating pyrophosphate (PP). Interestingly enough, the new DNA contains the same base ratio as the primer, which suggests that the existing DNA acted in some way to determine the chemical sequence. Although no one as yet has been able to make DNA with known biological properties, present experimental results strongly support the conclusion that DNA acts as a template for the replication of a duplicate of itself. From what we have said, it is clear that this duplication depends significantly on the energy metabolism of the cell, and, as we will discuss later, these cellular functions are in turn regulated by the DNA.

The basic molecular structure proposed for DNA has raised a number of questions in connection with the structure of the chromosome

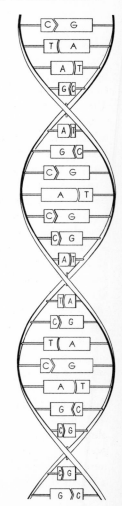

itself. Important experiments on chromosome and DNA duplication are discussed in another volume in this series,[3] and the reader is referred to other advanced books for a theoretical discussion of genetic coding and of the functioning of the base sequence in determining specific protein structure.

The Function of DNA

Probably the best direct evidence of DNA's role as a hereditary determinant in metabolism comes from studies of the *transformation* of certain bacteria, particularly pneumococcia. These experiments have demonstrated that the DNA from one strain of bacteria can alter the inherited metabolic capacities of a second strain. For example, by mutational techniques a strain of pneumococcus that is incapable of utilizing mannitol as a substrate can be isolated, and these bacterial cells do not contain the enzyme, mannitol phosphate dehydrogenase (M^- cells). When DNA is isolated from M^+ cells and added to a culture of M^- cells, many of the M^- cells are transformed into M^+ cells. The progeny of these cells are also capable of making the enzyme, thus indicating quite convincingly that the hereditary capacity of the cells is permanently altered. These results suggest that at least part of the genetic material in DNA is incorporated into the chromosomal structure of the host, and in a certain percentage of the cases this material is replicated during subsequent generations. How this DNA is incorporated into a permanent functional structure we do not know.

A number of experiments involving the transformation of genetic characters have been made, and we are now certain that the transforming agent is DNA. These experiments also strongly support the theory that DNA is the major chemical responsible for controlling the synthesis of the enzymic machinery from one generation to another.

The identification of DNA as the genetic messenger is also supported by new experiments using virus. From recent investigations of bacteriophage (bacterial virus), we know that these parasites attach themselves to the bacterial cell by means of their protein coat and that the DNA of the phage is immediately injected into the cell. Inside the bacterial cell, the phage DNA takes over the metabolic machinery and begins to make phage protein and new phage DNA. Eventually, the protein combines with the DNA to form a completed phage, and all this occurs at the metabolic expense of the host cell. The life cycle of the bacteriophage is schematically presented in Fig. 63. The evidence is quite clear that a "foreign" DNA is capable of influencing the synthetic capacities of cells. This fact is largely responsible for our present belief that an altered DNA or a virus DNA may be the cause of a number of diseases, particularly cancer.

---

[3] C. P. Swanson, *The Cell* (Englewood Cliffs, N.J.: Prentice-Hall, 1960).

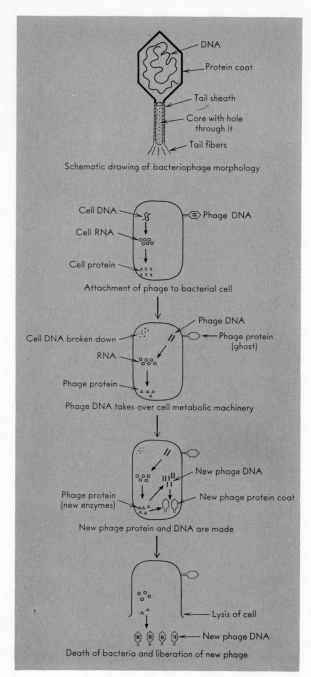

DNA

Protein coat

Tail sheath

Core with hole
through it

Tail fibers

Schematic drawing of bacteriophage morphology

Cell DNA

Cell RNA

Cell protein

Phage DNA

Attachment of phage to bacterial cell

Phage DNA

Cell DNA broken down

Phage protein
(ghost)

RNA

Phage protein

Phage DNA takes over cell metabolic machinery

New phage DNA

Phage protein
(new enzymes)

New phage protein coat

New phage protein and DNA are made

Lysis of cell

New phage DNA

Death of bacteria and liberation of new phage

Fig. 63. Life cycle of bacteriophage (T phage of *E. coli*).

ACTION OF NUCLEIC ACIDS IN PROTEIN SYNTHESIS

From what we now know about DNA, we can see that it is capable of inducing the formation of specific proteins, although there are numerous cases in which some protein synthesis can take place in the absence of DNA. The best working hypothesis at the present time is probably that of Caspersson, who has suggested that in some way the DNA directs the specific synthesis of nuclear proteins, which then transfer information to the synthesis of another type of nucleic acid, the ribose nucleic acid (RNA). In other words, the nuclear protein is fashioned to specifications dictated by the nucleotide sequences in DNA, and these proteins in turn command that the synthesis of RNA contain a specific sequence of bases. The RNA then moves into the cytoplasm of the cell to direct the biosynthesis of specific proteins. (Ninety per cent of cellular RNA is found in the cytoplasm.) Although there is a great deal of evidence to back up Caspersson's hypothesis, before presenting it we must discuss briefly the chemistry of RNA.

We know less about the macromolecular structure of RNA than we do about that of DNA. Important chemical differences between the two, however, have long been known. The five-carbon sugar attached to the bases, for example, is not deoxyribose but ribose. The pyrimidine base, thymine, is not found in RNA but is replaced by the base, uracil (Fig. 64); the four bases that make up RNA, therefore, are adenine, guanine, cytosine, and uracil. RNA is also a helical structure, and its nucleotides are spaced in much the same way as those in DNA. Ochoa and his associates, employing enzymatic methods, were recently able to synthesize a macromolecule from the nucleotide diphosphates that have many of the properties of RNA. Interestingly, the enzyme preparation that catalyzed the condensation reaction of the nucleotides will work with one of the bases to form a polynucleotide. If we add only ADP to the enzyme, therefore, a polymer is formed which contains only the purine base, adenine (Poly A).

At present it is believed that RNA is first made in the nucleolus structure of the nucleus and then moves into the cytoplasm to become associated with proteins in the various cellular structures. That RNA directly participates in the synthesis of cytoplasmic proteins is fairly well supported by recent experiments on cell-free preparations, and from earlier research which (1) demonstrated that labeled amino acids are more rapidly incorporated into proteins that are associated with structures containing a great deal of RNA and (2) indicated a direct relation between the RNA content of the cell and the rate of protein synthesis.

Fig. 64. The structure of uracil.

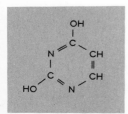

It is interesting to note that the small cellular particles called *microsomes* or ribosomes, which are associated with the membranous tubules and sac-like vesicles of the cytoplasm (generally called the endoplasmic reticulum) were the first to become labeled with the isotopic amino acid. Recently it has been possible to isolate the microsomes of certain cells and to demonstrate the synthesis of specific proteins by these particles, provided certain cytoplasmic proteins and an energy source is added along with the amino acids. (The synthesis of cytochrome C in cell-free preparations has also been observed, but this occurs in the mitochondria instead of in the microsomes.)

Details of cell-free protein synthesis are meager, but it now appears that a cytoplasmic soluble, low-molecular-weight RNA (10,000–15,000) acts as a transport system for active amino acids. As shown in Fig. 65, the initial step in the activation of amino acids for protein synthesis involves the formation of an energy-rich adenylic acid derivative of the amino acid. ATP has been shown to be essential for this reaction, and a number of different amino acid activating enzymes have now been purified and studied. Amino acids are activated in essentially the same way that fatty acids are activated (the latter process, you will recall, we discussed in a previous chapter).

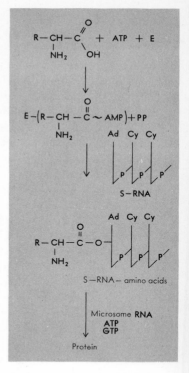

Fig. 65. Activation of amino acids and protein synthesis.

Later studies have shown that the amino acid adenylate transfers the amino acid to a specific soluble RNA (S-RNA), and present evidence indicates that the amino acid is attached to the second or third carbon of the ribose molecule in the terminal base of the S-RNA. The terminal sequence in S-RNA has in all cases been shown to be adenylic–cytidylic–cytidylic–, and the amino acid is attached to the adenylic acid. The S-RNA amino acids are in some way transferred to the microsomal RNA, where protein synthesis occurs. Very little is known about this last step, except that ATP and catalytic amounts of guanosine triphosphate are required. Presumably, the specific base pairing of S-RNA amino acids occurs on the RNA of the microsome in such a way that genetic information is transported to the amino acid sequence leading up to the formation of a protein. These events are schematically presented in Fig. 66.

We know enough today to hypothesize that RNA is directly concerned in the final fabrication of proteins from amino acids, but the details of the mechanisms involved are not understood. A number of interesting questions remain to be answered. Are there separate RNA templates for each protein? Do the various microsomal particles contain the same RNA information? Are the proteins synthesized in the nucleus,

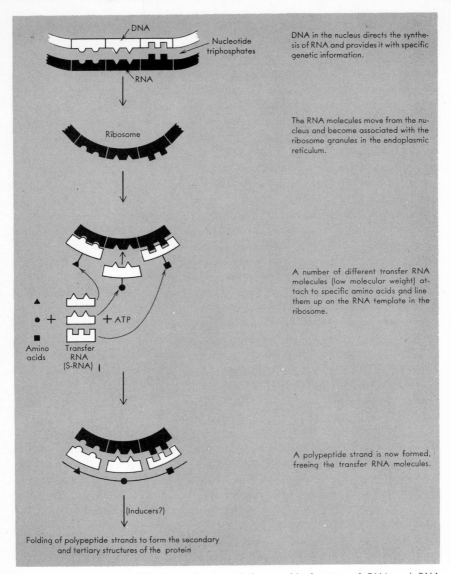

Fig. 66. Schematic diagram of the possible function of DNA and RNA in protein synthesis.

and dependent on DNA rather than on RNA proteins that are primarily concerned with RNA synthesis?

From what we have said about induced enzyme formation, it is clear that a number of factors other than the ones mentioned above are important in controlling protein synthesis. One RNA template may conceivably act as the site for the formation of a number of proteins. Specific inducers may influence the RNA-polypeptide site if their effect is primarily on the secondary and tertiary structures of a protein. Many proteins may have the same amino acid sequence but have differences in the size and shape of their structures, so that they have different enzymatic specifications. We saw earlier that the amino acid sequence in a number of enzymes could be the same and the enzymatic activities could still be quite different. In other words, the repeating patterns in protein synthesis suggest that a common code exists for at least parts of the molecule.

Some of the most valuable evidence in support of the common template idea comes from studies of antibody synthesis. If an animal, such as a rabbit, is injected repeatedly with a solution of a foreign protein, after a time the blood will contain specific antibodies to the protein administered. The administered protein is called the antigen. If a solution of the antigen is mixed with a sample of the serum, the antigen and antibody will combine and under appropriate conditions will form a precipitate. The specificity of this type of serological test is often so great that we can differentiate between preparations of a given protein from different species. Many efforts have been made to determine the chemical differences among antibodies that are specific for a given antigen, but all indications are that antibodies are indistinguishable from each other and from the proteins of the gamma globulin fraction of plasma from which they come. Antibodies to a large number of different antigens have been shown to have the same particle weight, the same amino acid composition, and the same terminal amino acids as do the normal gamma globulins.

These findings support the hypothesis that normal and immune gamma globulins are formed at the same site and that the differences are to be found not in the amino acid sequence but in the secondary structure. Specific RNA structures, therefore, are probably important in the building of large polypeptide chains, but these chains can unfold in a number of ways to form specific configurations which endow them with particular catalytic activity and/or specificity. Thus the surface configurations of the molecules of proteins, RNA, and DNA may be crucial factors in controlling the function of the cell. If the direction of the unfolding of the polypeptide chain from the RNA template is the main cause of the differences in proteins, we have another example of the essential economy of nature, for nature, then, does not need a large number of different mechanisms to synthesize a polypeptide chain, but only a directional force to give it a new shape or form when it comes off the surface. This may be the real secret of why nature has made the catalyst such a large molecule.

If this concept of enzymes is correct, it may be possible to control enzyme activity as well as enzyme synthesis in the cell by the enzymes' own internal chemistry. In other words, there may be competition at the enzyme-forming center which can be influenced by a number of factors in the environment; indeed, there is evidence to indicate that the products of the biosynthetic pathways can inhibit the synthesis of enzymes that are important in the early sequence of this biosynthesis. The products of enzyme reactions, therefore, may have what is called a feedback effect on the synthesis of enzymes that are vital in their own biosynthesis (enzyme repression).

Let us examine how this phenomenon affects the scheme presented in Fig. 59. In that example, we now know that the presence of excess histidine inhibits the synthesis of one of the enzymes concerned in the biosynthesis of one of the precursors. This type of cellular control of the enzyme-forming center may be of great significance in the regulation of cellular growth and differentiation.

In contrast to the possibility that a number of proteins can be synthesized through the use of common information from one RNA template, we must keep in mind the one gene-one enzyme concept introduced earlier. In fact, there is good genetic and biochemical information which suggests that a mutation is a localized change in the sequence of the DNA molecule or in its three-dimensional relationships and that this change is reflected in the amino acid sequence of the protein. This fact, however, should not exclude the possibility that gene mutation can also affect the secondary or tertiary structures of the protein. We know of instances in which the charge or thermostability of a protein can be changed by a gene mutation. Thus an organism may be able to adapt to a new environment, or to temporary changes in the environment, by processes more subtle than the complete resynthesis of new protein.

Unfortunately, it is not definitely known whether the amino acid sequence is also altered. In only a few cases has a change in the amino acid sequence been the primary factor that alters the physical properties of a protein. One of the best-known examples of this takes place in the hemoglobin of sickle-cell anemia. In this disease, which is inherited as a single gene difference, the hemoglobin is electrophoretically different from the normal. Analysis of the amino acid sequence reveals that one particular glutamic acid residue is replaced with valine, and this alteration is sufficient to account for the change in charge. Whether gene mutation can produce protein differences without a change in amino acid sequence must remain unknown until more experiments are performed.

Before ending our discussion of the genetic control of cellular processes, we must mention briefly the importance of the genetic control of membrane activity. We have already discussed the general problem of cellular permeability and the active transport of nutrients across the cell membrane. Quite recently it has been shown that the catalytic units in the cell membrane which are capable of mobilizing cellular energy for active accumulation of substrates are under genetic control. For example, we have known for some time that certain bacterial cells contain enzymes for the metabolism of specific compounds but that the cells are unable to use them because the compounds cannot pass through the cell membrane. Specific mutants have now been obtained that have the ability to accumulate these substrates, and, consequently, the cell is now able to metabolize them. These active transport systems in the membrane, called *permeases*, respond to environmental and genetic changes in much the same way as enzyme synthesis does. Thus the chemical composition of the cellular interior can be regulated by a change in the catalytic character of the cellular membrane. Changes in permease activity may be another crucial factor in the regulation of growth, differentiation, and cellular function.

Selected Readings

Fruton, J. S., and S. Simmonds, *General Biochemistry,* 2nd ed. New York: Wiley, 1958.

White, Abraham, Philip Handler, Emil L. Smith, and DeWitt Stetten, *Principles of Biochemistry,* 2nd ed. New York: McGraw-Hill, 1959.
Two outstanding advanced texts on biochemistry. Both are well documented with references to original research papers, review articles, and specialized textbooks and treatises.

Baldwin, E., *Dynamic Aspects of Biochemistry,* 3rd ed. New York: Cambridge University Press, 1957. An outstanding introduction to the concepts of cellular metabolism, with particular emphasis on the intermediary metabolism of carbohydrates, fats, amino acids, and nucleic acids.

Neilands, J. B., and Paul K. Stumpf, *Outlines of Enzyme Chemistry,* 2nd ed. New York: Wiley, 1958. An excellent elementary introduction to the field of enzyme chemistry.

Giese, A. C., *Cell Physiology.* Philadelphia: Saunders, 1957.

Davson, Hugh, *A Textbook of General Physiology,* 2nd ed. Boston: Little Brown, 1959.

Heilbrunn, L. V., *An Outline of General Physiology,* 3rd ed. Philadelphia: Saunders, 1952.
Three of the best textbooks on cellular and general physiology. They are especially good on such subjects as permeability of cell membranes, electrical conductivity, muscle contraction, photo-biological processes, and the effect of the environment on cell growth, function, and metabolism.

Anfinsen, C. B., *The Molecular Basis of Evolution.* New York: Wiley, 1959. A stimulating discussion of the relationship between genes, nucleic acids, and protein structure that serves as a basis for understanding evolutionary processes.

Index

A

B

C

117